IN ASSOCIATION WITH

SQA

HODDER GIBSON
Model Papers
WITH ANSWERS

PLUS: Official SQA Specimen Paper
With Answers

Advanced Higher for CfE
Mathematics

2015 Specimen Question Paper
& Model Papers

HODDER GIBSON
AN HACHETTE UK COMPANY

This book contains the official 2015 SQA Specimen Question Paper for Advanced Higher for CfE Maths, with associated SQA approved answers modified from the official marking instructions that accompany the paper.

In addition the book contains model papers, together with answers, plus study skills advice. These papers, some of which may include a limited number of previously published SQA questions, have been specially commissioned by Hodder Gibson, and have been written by experienced senior teachers and examiners in line with the new Advanced Higher for CfE syllabus and assessment outlines, Spring 2015. This is not SQA material but has been devised to provide further practice for Advanced Higher for CfE examinations in 2016 and beyond.

Hodder Gibson is grateful to the copyright holders, as credited on the final page of the Answer Section, for permission to use their material. Every effort has been made to trace the copyright holders and to obtain their permission for the use of copyright material. Hodder Gibson will be happy to receive information allowing us to rectify any error or omission in future editions.

Hachette UK's policy is to use papers that are natural, renewable and recyclable products and made from wood grown in sustainable forests. The logging and manufacturing processes are expected to conform to the environmental regulations of the country of origin.

Orders: please contact Bookpoint Ltd, 130 Park Drive, Milton Park, Abingdon, Oxon OX14 4SE. Telephone: (44) 01235 827720. Fax: (44) 01235 400454. Lines are open 9.00–5.00, Monday to Saturday, with a 24-hour message answering service. Visit our website at www.hoddereducation.co.uk. Hodder Gibson can be contacted direct on: Tel: 0141 848 1609; Fax: 0141 889 6315; email: hoddergibson@hodder.co.uk

This collection first published in 2015 by
Hodder Gibson, an imprint of Hodder Education,
An Hachette UK Company
2a Christie Street
Paisley PA1 1NB

Typeset by Aptara, Inc.

Printed in the UK

A catalogue record for this title is available from the British Library

ISBN: 978-1-4718-6048-5

3 2 1

2016 2015

Introduction

Study Skills – what you need to know to pass exams!

Pause for thought

Many students might skip quickly through a page like this. After all, we all know how to revise. Do you really though?

Think about this:

"IF YOU ALWAYS DO WHAT YOU ALWAYS DO, YOU WILL ALWAYS GET WHAT YOU HAVE ALWAYS GOT."

Do you like the grades you get? Do you want to do better? If you get full marks in your assessment, then that's great! Change nothing! This section is just to help you get that little bit better than you already are.

There are two main parts to the advice on offer here. The first part highlights fairly obvious things but which are also very important. The second part makes suggestions about revision that you might not have thought about but which WILL help you.

Part 1

DOH! It's so obvious but …

Start revising in good time

Don't leave it until the last minute – this will make you panic.

Make a revision timetable that sets out work time AND play time.

Sleep and eat!

Obvious really, and very helpful. Avoid arguments or stressful things too – even games that wind you up. You need to be fit, awake and focused!

Know your place!

Make sure you know exactly **WHEN and WHERE** your exams are.

Know your enemy!

Make sure you know what to expect in the exam.

How is the paper structured?

How much time is there for each question?

What types of question are involved?

Which topics seem to come up time and time again?

Which topics are your strongest and which are your weakest?

Are all topics compulsory or are there choices?

Learn by DOING!

There is no substitute for past papers and practice papers – they are simply essential! Tackling this collection of papers and answers is exactly the right thing to be doing as your exams approach.

Part 2

People learn in different ways. Some like low light, some bright. Some like early morning, some like evening / night. Some prefer warm, some prefer cold. But everyone uses their BRAIN and the brain works when it is active. Passive learning – sitting gazing at notes – is the most INEFFICIENT way to learn anything. Below you will find tips and ideas for making your revision more effective and maybe even more enjoyable. What follows gets your brain active, and active learning works!

Activity 1 – Stop and review

Step 1

When you have done no more than 5 minutes of revision reading STOP!

Step 2

Write a heading in your own words which sums up the topic you have been revising.

Step 3

Write a summary of what you have revised in no more than two sentences. Don't fool yourself by saying, "I know it, but I cannot put it into words". That just means you don't know it well enough. If you cannot write your summary, revise that section again, knowing that you must write a summary at the end of it. Many of you will have notebooks full of blue/black ink writing. Many of the pages will not be especially attractive or memorable so try to liven them up a bit with colour as you are reviewing and rewriting. **This is a great memory aid, and memory is the most important thing.**

Activity 2 – Use technology!

Why should everything be written down? Have you thought about "mental" maps, diagrams, cartoons and colour to help you learn? And rather than write down notes, why not record your revision material?

What about having a text message revision session with friends? Keep in touch with them to find out how and what they are revising and share ideas and questions.

Why not make a video diary where you tell the camera what you are doing, what you think you have learned and what you still have to do? No one has to see or hear it, but the process of having to organise your thoughts in a formal way to explain something is a very important learning practice.

Be sure to make use of electronic files. You could begin to summarise your class notes. Your typing might be slow, but it will get faster and the typed notes will be easier to read than the scribbles in your class notes. Try to add different fonts and colours to make your work stand out. You can easily Google relevant pictures, cartoons and diagrams which you can copy and paste to make your work more attractive and **MEMORABLE**.

Activity 3 – This is it. Do this and you will know lots!

Step 1

In this task you must be very honest with yourself! Find the SQA syllabus for your subject (www.sqa.org.uk). Look at how it is broken down into main topics called MANDATORY knowledge. That means stuff you MUST know.

Step 2

BEFORE you do ANY revision on this topic, write a list of everything that you already know about the subject. It might be quite a long list but you only need to write it once. It shows you all the information that is already in your long-term memory so you know what parts you do not need to revise!

Step 3

Pick a chapter or section from your book or revision notes. Choose a fairly large section or a whole chapter to get the most out of this activity.

With a buddy, use Skype, Facetime, Twitter or any other communication you have, to play the game "If this is the answer, what is the question?". For example, if you are revising Geography and the answer you provide is "meander", your buddy would have to make up a question like "What is the word that describes a feature of a river where it flows slowly and bends often from side to side?".

Make up 10 "answers" based on the content of the chapter or section you are using. Give this to your buddy to solve while you solve theirs.

Step 4

Construct a wordsearch of at least 10 × 10 squares. You can make it as big as you like but keep it realistic. Work together with a group of friends. Many apps allow you to make wordsearch puzzles online. The words and phrases can go in any direction and phrases can be split. Your puzzle must only contain facts linked to the topic you are revising. Your task is to find 10 bits of information to hide in your puzzle, but you must not repeat information that you used in Step 3. DO NOT show where the words are. Fill up empty squares with random letters. Remember to keep a note of where your answers are hidden but do not show your friends. When you have a complete puzzle, exchange it with a friend to solve each other's puzzle.

Step 5

Now make up 10 questions (not "answers" this time) based on the same chapter used in the previous two tasks. Again, you must find NEW information that you have not yet used. Now it's getting hard to find that new information! Again, give your questions to a friend to answer.

Step 6

As you have been doing the puzzles, your brain has been actively searching for new information. Now write a NEW LIST that contains only the new information you have discovered when doing the puzzles. Your new list is the one to look at repeatedly for short bursts over the next few days. Try to remember more and more of it without looking at it. After a few days, you should be able to add words from your second list to your first list as you increase the information in your long-term memory.

FINALLY! Be inspired...

Make a list of different revision ideas and beside each one write **THINGS I HAVE** tried, **THINGS I WILL** try and **THINGS I MIGHT** try. Don't be scared of trying something new.

And remember – "FAIL TO PREPARE AND PREPARE TO FAIL!"

Advanced Higher for CfE Mathematics

The course

The Advanced Higher Mathematics course is designed to build upon and extend the skills, knowledge and understanding that you have attained in the Higher Mathematics course (or equivalent qualification). It enables you to develop further skills in calculus, algebra and geometry. Areas such as number theory, complex numbers and matrices are introduced as well as processes of rigorous proof.

How the course is assessed

To gain the course award, you must pass the three Units:

- Methods in Algebra and Calculus
- Applications of Algebra and Calculus
- Geometry, Proof and Systems of Equations

as well as the examination.

The Units are assessed internally on a pass/fail basis.

The examination is set and marked by experienced practitioners appointed by SQA.

The course award is graded A–D, the grade being determined by the total mark you score in the examination.

The examination

The examination is a three hour paper with a total of 100 marks, in which the use of a calculator is permitted. A formulae list will be provided (see page two of the specimen question paper).

The question paper consists of short- and extended-response questions that require the application of skills developed in the course. You are expected to communicate responses clearly and to justify solutions.

Further details can be found in the Advanced Higher Mathematics section on the SQA website: www.sqa.org.uk/sqa/48507.html.

Key tips for your success

Practise! Practise! Practise!

DOING maths questions is the most effective use of your study time. You will benefit much more from spending 30 minutes doing maths questions than spending several hours copying out notes or reading a maths textbook. Practise basic skills such as the product and quotient rules regularly.

Prior learning

Ensure that you know trigonometric identities and other relevant formulae from the Higher Mathematics course, as well as essential basic techniques such as solving quadratic equations.

Show all working clearly

The instructions on the front of the exam paper state that "Full credit will be given only to solutions which contain appropriate working." A "correct" answer with no working may only be awarded partial marks or even no marks at all. An incomplete answer will be awarded marks for any appropriate working.

Attempt every question, even if you are not sure whether you are correct or not. Your solution may contain working which will gain some marks. A blank response is certain to be awarded no marks.

Never cross out working unless you have something better to replace it with.

Ensure that you communicate reasons for what you have done, wherever appropriate. In particular, in proof and "show that" questions, include all lines of working.

Marking instructions

Ensure that you look at the detailed marking instructions of model papers and past papers. They provide further advice and guidelines as well as showing you precisely where, and for what, marks are awarded.

Extended response questions

You should look for connections between parts of questions, particularly where there are three or four sections to a question. These are almost always linked and, in some instances, an earlier result in part (a) or (b) is needed and its use would avoid further repeated work.

Accuracy

Where possible use exact values; decimal approximations may lead to inaccuracies which could cost you marks.

Notation

In all questions make sure that you use the correct notation. In particular, for integration questions, remember to include 'dx' within your integral. When finding a definite integral, remember to include the constant of integration in your answer.

Radians

Remember to work in radians when attempting any question involving both trigonometry and calculus.

Simplify

Get into the habit of simplifying expressions before doing any further work with them. This should make all subsequent work easier.

Good luck!

Remember that the rewards for passing Advanced Higher Mathematics are well worth it! Your pass will help you get the future you want for yourself. In the exam, be confident in your own ability; if you're not sure how to answer a question, trust your instincts and give it a go anyway – keep calm and don't panic! GOOD LUCK!

ADVANCED HIGHER FOR CfE

2015 Specimen Question Paper

National Qualifications
SPECIMEN ONLY

Mathematics

Duration — 3 hours

Total marks — 100

Attempt ALL questions.

You may use a calculator.

Full credit will be given only to solutions which contain appropriate working.

State the units for your answer where appropriate.

Write your answers clearly in the answer booklet provided. In the answer booklet, you must clearly identify the question number you are attempting

Use **blue** or **black** ink.

Before leaving the examination room you must give your answer booklet to the Invigilator; if you do not, you may lose all the marks for this paper.

FORMULAE LIST

Standard derivatives		Standard integrals	
$f(x)$	$f'(x)$	$f(x)$	$\int f(x)\,dx$
$\sin^{-1}x$	$\dfrac{1}{\sqrt{1-x^2}}$	$\sec^2(ax)$	$\dfrac{1}{a}\tan(ax)+c$
$\cos^{-1}x$	$-\dfrac{1}{\sqrt{1-x^2}}$	$\dfrac{1}{\sqrt{a^2-x^2}}$	$\sin^{-1}\left(\dfrac{x}{a}\right)+c$
$\tan^{-1}x$	$\dfrac{1}{1+x^2}$	$\dfrac{1}{a^2+x^2}$	$\dfrac{1}{a}\tan^{-1}\left(\dfrac{x}{a}\right)+c$
$\tan x$	$\sec^2 x$	$\dfrac{1}{x}$	$\ln|x|+c$
$\cot x$	$-\operatorname{cosec}^2 x$	e^{ax}	$\dfrac{1}{a}e^{ax}+c$
$\sec x$	$\sec x \tan x$		
$\operatorname{cosec} x$	$-\operatorname{cosec} x \cot x$		
$\ln x$	$\dfrac{1}{x}$		
e^x	e^x		

Summations

(Arithmetic series) $S_n = \dfrac{1}{2}n[2a+(n-1)d]$

(Geometric series) $S_n = \dfrac{a(1-r^n)}{1-r}$

$$\sum_{r=1}^{n} r = \frac{n(n+1)}{2}, \quad \sum_{r=1}^{n} r^2 = \frac{n(n+1)(2n+1)}{6}, \quad \sum_{r=1}^{n} r^3 = \frac{n^2(n+1)^2}{4}$$

Binomial theorem

$$(a+b)^n = \sum_{r=0}^{n} \binom{n}{r} a^{n-r}b^r \quad \text{where} \quad \binom{n}{r} = {}^nC_r = \frac{n!}{r!(n-r)!}$$

Maclaurin expansion

$$f(x) = f(0) + f'(0)x + \frac{f''(0)x^2}{2!} + \frac{f'''(0)x^3}{3!} + \frac{f^{iv}(0)x^4}{4!} + \ldots$$

De Moivre's theorem

$$[r(\cos\theta + i\sin\theta)]^n = r^n(\cos n\theta + i\sin n\theta)$$

Vector product

$$\mathbf{a}\times\mathbf{b} = |\mathbf{a}||\mathbf{b}|\sin\theta\,\hat{\mathbf{n}} = \begin{vmatrix} \mathbf{i} & \mathbf{j} & \mathbf{k} \\ a_1 & a_2 & a_3 \\ b_1 & b_2 & b_3 \end{vmatrix} = \mathbf{i}\begin{vmatrix} a_2 & a_3 \\ b_2 & b_3 \end{vmatrix} - \mathbf{j}\begin{vmatrix} a_1 & a_3 \\ b_1 & b_3 \end{vmatrix} + \mathbf{k}\begin{vmatrix} a_1 & a_2 \\ b_1 & b_2 \end{vmatrix}$$

Matrix transformation

Anti-clockwise rotation through an angle, θ about the origin, $\begin{bmatrix} \cos\theta & -\sin\theta \\ \sin\theta & \cos\theta \end{bmatrix}$

$$[r(\cos\theta + i\sin\theta)]^n = r^n(\cos n\theta + i\sin n\theta)$$

Total marks — 100

MARKS

Attempt ALL questions

1. Given $f(x) = \dfrac{x-1}{1+x^2}$, show that $f'(x) = \dfrac{1+2x-x^2}{(1+x^2)^2}$.   **3**

2. State and simplify the general term in the binomial expansion of $\left(2x - \dfrac{5}{x^2}\right)^6$.

    Hence, or otherwise, find the term independent of x.   **3**

3. Find $\displaystyle\int \dfrac{2}{\sqrt{9-16x^2}}\,dx$.   **3**

4. Show that the greatest common divisor of 487 and 729 is 1.

    Hence find integers x and y such that $487x + 729y = 1$.   **4**

5. Find $\displaystyle\int x^2 e^{3x}\,dx$.   **5**

6. Find the values of the constant k for which the matrix $\begin{pmatrix} 3 & k & 2 \\ 3 & -4 & 2 \\ k & 0 & 1 \end{pmatrix}$ is singular.   **4**

7. A spherical balloon is being inflated. When the radius is 10 cm the surface area is increasing at a rate of $120\pi\,\text{cm}^2\,\text{s}^{-1}$.

    Find the rate at which the volume is increasing at this moment.   **5**

    (Volume of sphere $= \dfrac{4}{3}\pi r^3$, surface area $= 4\pi r^2$)

8. (a) Find the Maclaurin expansions up to and including the term in x^3, simplifying the coefficients as far as possible, for the following:

      (i) $f(x) = e^{3x}$

      (ii) $g(x) = (x+2)^{-2}$ **5**

    (b) Given that $h(x) = \dfrac{xe^{3x}}{(x+2)^2}$ use the expansions from (a) to approximate the value

      of $h\left(\dfrac{1}{2}\right)$.   **3**

MARKS

9. Three terms of an arithmetic sequence, u_3, u_7 and u_{16} form the first three terms of a geometric sequence.

Show that $a = \dfrac{6}{5}d$, where a and d are, respectively, the first term and common difference of the arithmetic sequence with $d \neq 0$.

Hence, or otherwise, find the value of r, the common ratio of the geometric sequence. **4**

10. Using logarithmic differentiation, or otherwise, find $\dfrac{dy}{dx}$ given that

$$e^y = \frac{(3x+2)e^{2x}}{(2x-1)^2}, \quad x > \frac{1}{2}.$$ **3**

11. Find the exact value of $\displaystyle\int_1^2 \frac{x+4}{(x+1)^2(2x-1)}\,dx$. **7**

12. (a) Given that m and n are positive integers state the negation of the statement:

m is even or n is even. **1**

(b) By considering the contrapositive of the following statement:

if mn is even then m is even or n is even,

prove that the statement is true for all positive integers m and n. **3**

13. Consider the curve in the (x, y) plane defined by the equation $y = \dfrac{4x-3}{x^2-2x-8}$.

(a) Identify the vertical asymptotes to this curve and justify your answer. **2**

Here are two statements about the curve:

(1) It does not cross or touch the x-axis.

(2) The line $y = 0$ is an asymptote.

(b) (i) State why statement (1) is false.

(ii) Show that statement (2) is true. **3**

MARKS

14. The lines L_1 and L_2 are given by the following equations:

$$L_1: \frac{x+6}{3} = \frac{y-1}{-1} = \frac{z-2}{2}$$

$$L_2: \frac{x+5}{4} = \frac{y+4}{1} = \frac{z}{4}$$

(a) Show that the lines L_1 and L_2 intersect and state the coordinates of the point of intersection.   5

(b) Find the equation of the plane containing L_1 and L_2.   3

A third line, L_3, is given by the equation $\dfrac{x-1}{2} = \dfrac{y+7}{4} = \dfrac{z-3}{-1}$.

(c) Calculate the acute angle between L_3 and the plane. Give your answer in degrees correct to 2 decimal places.   4

15. (a) Given that $f(x) = \ln\left(\dfrac{1+x}{1-x}\right)$, find $f'(x)$, expressing your answer as a single fraction.   2

(b) Solve the differential equation

$$\cos x \frac{dy}{dx} + y \tan x = \frac{\cos x}{e^{\sec x}}$$

given that $y = 1$ when $x = 2\pi$. Express your answer in the form $y = f(x)$.   7

16. Let $S_n = \displaystyle\sum_{r=1}^{n} \frac{1}{r(r+1)}$ where n is a positive integer.

(a) Prove that, for all positive integers n, $S_n = \dfrac{n}{n+1}$.   5

(b) Find

(i) the least value of n such that $S_{n+1} - S_n < \dfrac{1}{1000}$

(ii) the value of n for which $S_n \times S_{n-1} \times S_{n-2} = S_{n-8}$.   5

MARKS

17. (a) Given $z = \cos\theta + i\sin\theta$, use de Moivre's theorem and the binomial theorem to show that:

$$\cos 4\theta = \cos^4\theta - 6\cos^2\theta\sin^2\theta + \sin^4\theta$$

and

$$\sin 4\theta = 4\cos^3\theta\sin\theta - 4\cos\theta\sin^3\theta.$$

5

(b) Hence show that $\tan 4\theta = \dfrac{4\tan\theta - 4\tan^3\theta}{1 - 6\tan^2\theta + \tan^4\theta}$.

3

(c) Find algebraically the solutions to the equation

$$\tan^4\theta + 4\tan^3\theta - 6\tan^2\theta - 4\tan\theta + 1 = 0$$

in the interval $0 \le \theta \le \dfrac{\pi}{2}$.

3

[END OF SPECIMEN QUESTION PAPER]

[BLANK PAGE]

ADVANCED HIGHER FOR CfE

Model Paper 1

Whilst this Model Paper has been specially commissioned by Hodder Gibson for use as practice for the Advanced Higher (for Curriculum for Excellence) exams, the key reference document remains the SQA Specimen Paper 2015.

National
Qualifications
MODEL PAPER 1

Mathematics

Duration — 3 hours

Total marks — 100

Attempt ALL questions.

You may use a calculator.

Full credit will be given only to solutions which contain appropriate working.

State the units for your answer where appropriate.

Write your answers clearly in the answer booklet provided. In the answer booklet, you must clearly identify the question number you are attempting.

Use **blue** or **black** ink.

Before leaving the examination room you must give your answer booklet to the Invigilator; if you do not, you may lose all the marks for this paper.

FORMULAE LIST

Standard derivatives	
$f(x)$	$f'(x)$
$\sin^{-1}x$	$\dfrac{1}{\sqrt{1-x^2}}$
$\cos^{-1}x$	$-\dfrac{1}{\sqrt{1-x^2}}$
$\tan^{-1}x$	$\dfrac{1}{1+x^2}$
$\tan x$	$\sec^2 x$
$\cot x$	$-\operatorname{cosec}^2 x$
$\sec x$	$\sec x \tan x$
$\operatorname{cosec} x$	$-\operatorname{cosec} x \cot x$
$\ln x$	$\dfrac{1}{x}$
e^x	e^x

Standard integrals	
$f(x)$	$\int f(x)\,dx$
$\sec^2(ax)$	$\dfrac{1}{a}\tan(ax)+c$
$\dfrac{1}{\sqrt{a^2-x^2}}$	$\sin^{-1}\left(\dfrac{x}{a}\right)+c$
$\dfrac{1}{a^2+x^2}$	$\dfrac{1}{a}\tan^{-1}\left(\dfrac{x}{a}\right)+c$
$\dfrac{1}{x}$	$\ln|x|+c$
e^{ax}	$\dfrac{1}{a}e^{ax}+c$

Summations

(Arithmetic series) $S_n = \dfrac{1}{2}n[2a+(n-1)d]$

(Geometric series) $S_n = \dfrac{a(1-r^n)}{1-r}$

$$\sum_{r=1}^{n} r = \frac{n(n+1)}{2}, \quad \sum_{r=1}^{n} r^2 = \frac{n(n+1)(2n+1)}{6}, \quad \sum_{r=1}^{n} r^3 = \frac{n^2(n+1)^2}{4}$$

Binomial theorem

$$(a+b)^n = \sum_{r=0}^{n} \binom{n}{r} a^{n-r} b^r \quad \text{where} \quad \binom{n}{r} = {}^nC_r = \frac{n!}{r!(n-r)!}$$

Maclaurin expansion

$$f(x) = f(0) + f'(0)x + \frac{f''(0)x^2}{2!} + \frac{f'''(0)x^3}{3!} + \frac{f^{iv}(0)x^4}{4!} + \ldots$$

De Moivre's theorem

$$[r(\cos\theta + i\sin\theta)]^n = r^n(\cos n\theta + i\sin n\theta)$$

Vector product

$$\mathbf{a}\times\mathbf{b} = |\mathbf{a}||\mathbf{b}|\sin\theta\,\hat{\mathbf{n}} = \begin{vmatrix} \mathbf{i} & \mathbf{j} & \mathbf{k} \\ a_1 & a_2 & a_3 \\ b_1 & b_2 & b_3 \end{vmatrix} = \mathbf{i}\begin{vmatrix} a_2 & a_3 \\ b_2 & b_3 \end{vmatrix} - \mathbf{j}\begin{vmatrix} a_1 & a_3 \\ b_1 & b_3 \end{vmatrix} + \mathbf{k}\begin{vmatrix} a_1 & a_2 \\ b_1 & b_2 \end{vmatrix}$$

Matrix transformation

Anti-clockwise rotation through an angle, θ about the origin, $\begin{bmatrix} \cos\theta & -\sin\theta \\ \sin\theta & \cos\theta \end{bmatrix}$

Total marks — 100

Attempt ALL questions

MARKS

1.   (a)   Given $f(x) = (x + 1)(x - 2)^3$, obtain the values of x for which $f'(x) = 0$.   **3**

     (b)   Calculate the gradient of the curve defined by $\dfrac{x^2}{y} + x = y - 5$ at the point $(3, -1)$.   **4**

2.   The first term of an arithmetic sequence is 2 and the 20th term is 97. Obtain the sum of the first 50 terms.   **4**

3.   Show that $z = 3 + 3i$ is a root of the equation $z^3 - 18z + 108 = 0$ and obtain the remaining roots of the equation.   **4**

4.   Let the matrix $A = \begin{pmatrix} 1 & x \\ x & 4 \end{pmatrix}$.

     (a)   Obtain the value(s) of x for which A is singular.   **2**

     (b)   When $x = 2$, show that $A^2 = pA$ for some constant p.

       Determine the value of q such that $A^4 = qA$.   **3**

5.   (a)   Write down the binomial expansion of $(1 + x)^5$.   **1**

     (b)   Hence show that $0 \cdot 9^5$ is $0 \cdot 59049$.   **2**

6.   Use the substitution $x = 1 + \sin\theta$ to evaluate $\displaystyle\int_0^{\frac{\pi}{2}} \dfrac{\cos\theta}{(1 + \sin\theta)^3}\, d\theta$.   **5**

7.   Obtain the first three non-zero terms in the Maclaurin expansion of $(1 + \sin^2 x)$.   **4**

8.   Prove by induction that, for all positive integers n,

$$\sum_{r=1}^{n} \frac{1}{r(r+1)} = 1 - \frac{1}{n+1}.$$

   5

MARKS

9. Given that $y > -1$ and $x > -1$, obtain the general solution of the differential equation

$$\frac{dy}{dx} = 3(1+y)\sqrt{1+x},$$

expressing your answer in the form $y = f(x)$.   **5**

10. Use integration by parts to obtain the exact value of $\int_0^1 x\tan^{-1}x^2\,dx$.   **5**

11. A body moves along a straight line with velocity $v = t^3 - 12t^2 + 32t$ at time t.

    (a) Obtain the value of its acceleration when $t = 0$.   **1**

    (b) At time $t = 0$, the body is at the origin O.

    Obtain a formula for the displacement of the body at time t.

    Show that the body returns to O, and obtain the time, T, when this happens.   **4**

12. Given that $|z - 2| = |z + i|$, where $z = x + iy$, show that $ax + by + c = 0$ for suitable values of a, b and c.

    Indicate on an Argand diagram the locus of complex numbers z which satisfy $|z - 2| = |z + i|$.   **4**

13. Prove by contradiction that if x is an irrational number, then $2 + x$ is irrational.   **4**

14. Obtain the general solution of the differential equation

$$\frac{d^2y}{dx^2} - 3\frac{dy}{dx} + 2y = 2x^2.$$

Given that $y = \frac{1}{2}$ and $\frac{dy}{dx} = 1$ when $x = 0$, find the particular solution.   **10**

MARKS

15. Express $\dfrac{1}{x^3 + x}$ in partial fractions.

Obtain a formula for $I(k)$, where $I(k) = \displaystyle\int_1^k \dfrac{1}{x^3 + x}\, dx$, expressing it in the form $\ln\dfrac{a}{b}$, where a and b depend on k.

Write down an expression for $e^{I(k)}$ and obtain the value of $\lim_{k \to \infty} e^{I(k)}$. 　　10

16. Let $f(x) = \dfrac{x}{\ln x}$ for $x > 1$.

　(a) Derive expressions for $f'(x)$ and $f''(x)$, simplifying your answers. 　　4

　(b) Obtain the coordinates and nature of the stationary point of the curve $y = f(x)$. 　　3

　(c) Obtain the coordinates of the point of inflexion. 　　2

17. (a) Use Gaussian elimination on the following system of equations to give an expression for z in terms of λ.

$$x + y - z = 6$$
$$2x - 3y + 2z = 2$$
$$-5x + 2y + 4\lambda = 1$$

Determine the solution to this system of equations when $\lambda = -4$. 　　5

　(b) Show that the line of intersection, L, of the planes $x + y - z = 6$ and $2x - 3y + 2z = 2$ has parametric equations

$$x = t$$
$$y = 4t - 14$$
$$z = 5t - 20.$$ 　　2

　(c) Find the acute angle between line L and the plane $-5x + 2y - 4z = 1$. 　　4

[END OF MODEL PAPER]

National
Qualifications
MODEL PAPER 2

Mathematics

Duration — 3 hours

Total marks — 100

Attempt ALL questions.

You may use a calculator.

Full credit will be given only to solutions which contain appropriate working.

State the units for your answer where appropriate.

Write your answers clearly in the answer booklet provided. In the answer booklet, you must clearly identify the question number you are attempting.

Use **blue** or **black** ink.

Before leaving the examination room you must give your answer booklet to the Invigilator; if you do not, you may lose all the marks for this paper.

HODDER
GIBSON
LEARN MORE

FORMULAE LIST

Standard derivatives	
$f(x)$	$f'(x)$
$\sin^{-1}x$	$\dfrac{1}{\sqrt{1-x^2}}$
$\cos^{-1}x$	$-\dfrac{1}{\sqrt{1-x^2}}$
$\tan^{-1}x$	$\dfrac{1}{1+x^2}$
$\tan x$	$\sec^2 x$
$\cot x$	$-\mathrm{cosec}^2 x$
$\sec x$	$\sec x \tan x$
$\mathrm{cosec}\,x$	$-\mathrm{cosec}\,x \cot x$
$\ln x$	$\dfrac{1}{x}$
e^x	e^x

Standard integrals	
$f(x)$	$\int f(x)\,dx$
$\sec^2(ax)$	$\dfrac{1}{a}\tan(ax)+c$
$\dfrac{1}{\sqrt{a^2-x^2}}$	$\sin^{-1}\left(\dfrac{x}{a}\right)+c$
$\dfrac{1}{a^2+x^2}$	$\dfrac{1}{a}\tan^{-1}\left(\dfrac{x}{a}\right)+c$
$\dfrac{1}{x}$	$\ln|x|+c$
e^{ax}	$\dfrac{1}{a}e^{ax}+c$

Summations

(Arithmetic series) $\qquad S_n = \dfrac{1}{2}n[2a+(n-1)d]$

(Geometric series) $\qquad S_n = \dfrac{a(1-r^n)}{1-r}$

$$\sum_{r=1}^{n} r = \frac{n(n+1)}{2}, \quad \sum_{r=1}^{n} r^2 = \frac{n(n+1)(2n+1)}{6}, \quad \sum_{r=1}^{n} r^3 = \frac{n^2(n+1)^2}{4}$$

Binomial theorem

$$(a+b)^n = \sum_{r=0}^{n} \binom{n}{r} a^{n-r}b^r \quad \text{where} \quad \binom{n}{r} = {}^nC_r = \frac{n!}{r!(n-r)!}$$

Maclaurin expansion

$$f(x) = f(0) + f'(0)x + \frac{f''(0)x^2}{2!} + \frac{f'''(0)x^3}{3!} + \frac{f^{iv}(0)x^4}{4!} + \dots$$

De Moivre's theorem

$$[r(\cos\theta + i\sin\theta)]^n = r^n(\cos n\theta + i\sin n\theta)$$

Vector product

$$\mathbf{a} \times \mathbf{b} = |\mathbf{a}||\mathbf{b}|\sin\theta\,\hat{\mathbf{n}} = \begin{vmatrix} \mathbf{i} & \mathbf{j} & \mathbf{k} \\ a_1 & a_2 & a_3 \\ b_1 & b_2 & b_3 \end{vmatrix} = \mathbf{i}\begin{vmatrix} a_2 & a_3 \\ b_2 & b_3 \end{vmatrix} - \mathbf{j}\begin{vmatrix} a_1 & a_3 \\ b_1 & b_3 \end{vmatrix} + \mathbf{k}\begin{vmatrix} a_1 & a_2 \\ b_1 & b_2 \end{vmatrix}$$

Matrix transformation

Anti-clockwise rotation through an angle, θ about the origin, $\begin{bmatrix} \cos\theta & -\sin\theta \\ \sin\theta & \cos\theta \end{bmatrix}$

Total marks — 100   MARKS

Attempt ALL questions

1. Use the binomial theorem to expand $\left(\frac{1}{2}x - 3\right)^4$ and simplify your answer.   3

2. Differentiate, simplifying your answers:

    (a) $2\tan^{-1}\sqrt{1+x}$, where $x > -1$;   3

    (b) $\dfrac{1+\ln x}{3x}$, where $x > 0$.   3

3. The first and fourth terms of a geometric series are 2048 and 256 respectively. Calculate the value of the common ratio.

    Given that the sum of the first n terms is 4088, find the value of n.   5

4. Show that

    $$\int_{\ln\frac{3}{2}}^{\ln 2} \frac{e^x + e^{-x}}{e^x - e^{-x}}\, dx = \ln\frac{9}{5}.$$ 4

5. Express $z = \dfrac{(1+2i)^2}{7-i}$ in the form $a + bi$ where a and b are real numbers.

    Show z on an Argand diagram and evaluate $|z|$ and $\arg(z)$.   6

6. A curve is defined by the equation $xy^2 + 3x^2y = 4$ for $x > 0$ and $y > 0$.

    Use implicit differentiation to find $\dfrac{dy}{dx}$.

    Hence find an equation of the tangent to the curve where $x = 1$.   6

7. Use the substitution $u = 1 + x$ to evaluate $\displaystyle\int_0^3 \frac{x}{\sqrt{1+x}}\, dx$.   5

8. Use Gaussian elimination to solve the system of equations below when $\lambda \neq 2$:

    $$x + y + 2z = 1$$
    $$2x + \lambda y + z = 0$$
    $$3x + 3y + 9z = 5.$$

    Explain what happens when $\lambda = 2$.   6

MARKS

9. Given the matrix $A = \begin{pmatrix} 0 & 4 & 2 \\ 1 & 0 & 1 \\ -1 & -2 & -3 \end{pmatrix}$, show that $A^2 + A = kI$ for some constant k, where

I is the 3×3 unit matrix.

Obtain the values of p and q for which $A^{-1} = pA + qI$.   6

10. Show that $\sum_{r=1}^{n}(4 - 6r) = n - 3n^2$.

Hence write down a formula for $\sum_{r=1}^{2q}(4 - 6r)$.

Show that $\sum_{r=q+1}^{2q}(4 - 6r) = q - 9q^2$.   5

11. The amount x micrograms of an impurity removed per kg of a substance by a chemical process depends on the temperature $T°C$ as follows:

$x = T^3 - 90T^2 + 2400T, \quad 10 \leq T \leq 60.$

At what temperature in the given range should the process be carried out to remove as much impurity per kg as possible?   4

12. For each of the following statements, decide whether it is true or false and prove your conclusion.

(a) For all natural numbers m, if m^2 is divisible by 4 then m is divisible by 4.   2

(b) The cube of any odd integer p plus the square of any even integer q is always odd.   3

13. (a) Find an equation of the plane π_1 through the points A (1, 1, 1), B (2, −1, 1) and C (0, 3, 3).   3

(b) The plane π_2 has equation $x + 3y - z = 2$. Given that the point (0, a, b) lies on both the planes π_1 and π_2, find the values of a and b. Hence find an equation of the line of intersection of the planes π_1 and π_2.   4

(c) Find the size of the acute angle between the planes π_1 and π_2.   3

MARKS

14. Express $\dfrac{x^2+6x-4}{(x+2)^2(x-4)}$ in partial fractions.

Hence, or otherwise, obtain the first three non-zero terms in the Maclaurin

expansion of $\dfrac{x^2+6x-4}{(x+2)^2(x-4)}$.   9

15. A curve is defined by parametric equations $x=\cos 2t,\ y=\sin 2t,\ 0<t<\dfrac{\pi}{2}$.

   (a) Use parametric differentiation to find $\dfrac{dy}{dx}$.

       Hence find the equation of the tangent when $t=\dfrac{\pi}{8}$.   5

   (b) Obtain an expression for $\dfrac{d^2y}{dx^2}$ and hence show that $\sin 2t\,\dfrac{d^2y}{dx^2}+\left(\dfrac{dy}{dx}\right)^2=k$,

       where k is an integer. State the value of k.   5

16. A garden centre advertises young plants to be used as hedging. After planting, the growth G metres (i.e. the increase in height) after t years is modelled by the differential equation

$$\frac{dG}{dt}=\frac{25k-G}{25}$$

where k is a constant and $G=0$ when $t=0$.

   (a) Express G in terms of t and k.   4

   (b) Given that a plant grows 0·6 metres by the end of 5 years, find the value of k correct to three decimal places.   2

   (c) On the plant labels, it states that the expected growth after 10 years is approximately 1 metre. Is this claim justified?   2

   (d) Given that the initial height of the plants was 0·3 m, what is the likely long-term height of the plants?   2

[END OF MODEL PAPER]

ADVANCED HIGHER FOR CfE

Model Paper 3

Whilst this Model Paper has been specially commissioned by Hodder Gibson for use as practice for the Advanced Higher (for Curriculum for Excellence) exams, the key reference document remains the SQA Specimen Paper 2015.

National
Qualifications
MODEL PAPER 3

Mathematics

Duration — 3 hours

Total marks — 100

Attempt ALL questions.

You may use a calculator.

Full credit will be given only to solutions which contain appropriate working.

State the units for your answer where appropriate.

Write your answers clearly in the answer booklet provided. In the answer booklet, you must clearly identify the question number you are attempting.

Use blue or black ink.

Before leaving the examination room you must give your answer booklet to the Invigilator; if you do not, you may lose all the marks for this paper.

FORMULAE LIST

Standard derivatives	
$f(x)$	$f'(x)$
$\sin^{-1}x$	$\dfrac{1}{\sqrt{1-x^2}}$
$\cos^{-1}x$	$-\dfrac{1}{\sqrt{1-x^2}}$
$\tan^{-1}x$	$\dfrac{1}{1+x^2}$
$\tan x$	$\sec^2 x$
$\cot x$	$-\operatorname{cosec}^2 x$
$\sec x$	$\sec x \tan x$
$\operatorname{cosec} x$	$-\operatorname{cosec} x \cot x$
$\ln x$	$\dfrac{1}{x}$
e^x	e^x

Standard integrals	
$f(x)$	$\int f(x)\,dx$
$\sec^2(ax)$	$\dfrac{1}{a}\tan(ax)+c$
$\dfrac{1}{\sqrt{a^2-x^2}}$	$\sin^{-1}\left(\dfrac{x}{a}\right)+c$
$\dfrac{1}{a^2+x^2}$	$\dfrac{1}{a}\tan^{-1}\left(\dfrac{x}{a}\right)+c$
$\dfrac{1}{x}$	$\ln|x|+c$
e^{ax}	$\dfrac{1}{a}e^{ax}+c$

Summations

(Arithmetic series) $S_n = \dfrac{1}{2}n[2a+(n-1)d]$

(Geometric series) $S_n = \dfrac{a(1-r^n)}{1-r}$

$$\sum_{r=1}^{n} r = \frac{n(n+1)}{2}, \quad \sum_{r=1}^{n} r^2 = \frac{n(n+1)(2n+1)}{6}, \quad \sum_{r=1}^{n} r^3 = \frac{n^2(n+1)^2}{4}$$

Binomial theorem

$$(a+b)^n = \sum_{r=0}^{n} \binom{n}{r} a^{n-r} b^r \quad \text{where} \quad \binom{n}{r} = {}^nC_r = \frac{n!}{r!(n-r)!}$$

Maclaurin expansion

$$f(x) = f(0) + f'(0)x + \frac{f''(0)x^2}{2!} + \frac{f'''(0)x^3}{3!} + \frac{f^{iv}(0)x^4}{4!} + \ldots$$

De Moivre's theorem

$$[r(\cos\theta + i\sin\theta)]^n = r^n(\cos n\theta + i\sin n\theta)$$

Vector product

$$\mathbf{a}\times\mathbf{b} = |\mathbf{a}||\mathbf{b}|\sin\theta\,\hat{\mathbf{n}} = \begin{vmatrix} \mathbf{i} & \mathbf{j} & \mathbf{k} \\ a_1 & a_2 & a_3 \\ b_1 & b_2 & b_3 \end{vmatrix} = \mathbf{i}\begin{vmatrix} a_2 & a_3 \\ b_2 & b_3 \end{vmatrix} - \mathbf{j}\begin{vmatrix} a_1 & a_3 \\ b_1 & b_3 \end{vmatrix} + \mathbf{k}\begin{vmatrix} a_1 & a_2 \\ b_1 & b_2 \end{vmatrix}$$

Matrix transformation

Anti-clockwise rotation through an angle, θ about the origin, $\begin{bmatrix} \cos\theta & -\sin\theta \\ \sin\theta & \cos\theta \end{bmatrix}$

Total marks — 100   MARKS

Attempt ALL questions

1. The second and third terms of a geometric series are −6 and 3 respectively. Explain why the series has a sum to infinity, and obtain this sum.   5

2. (a) Differentiate $f(x) = \cos^{-1}(3x)$ where $-\dfrac{1}{3} < x < \dfrac{1}{3}$.   2

    (b) Given $x = 2\sec\theta$, $y = 3\sin\theta$, use parametric differentiation to find $\dfrac{dy}{dx}$ in terms of θ.   3

3. Given $\mathbf{u} = -2\mathbf{i} + 5\mathbf{k}$, $\mathbf{v} = 3\mathbf{i} + 2\mathbf{j} - \mathbf{k}$ and $\mathbf{w} = -\mathbf{i} + \mathbf{j} + 4\mathbf{k}$, calculate $\mathbf{u} \cdot (\mathbf{v} \times \mathbf{w})$.   4

4. Use the Euclidean algorithm to obtain the greatest common divisor of 1326 and 14654, expressing it in the form $1326a + 14654b$, where a and b are integers.   4

5. Matrices A and B are defined by

$$A = \begin{pmatrix} 1 & 0 & -1 \\ 0 & 1 & -1 \\ 0 & 1 & 2 \end{pmatrix}, \quad B = \begin{pmatrix} x+2 & x-2 & x+3 \\ -4 & 4 & 2 \\ 2 & -2 & 3 \end{pmatrix}.$$

    (a) Find the product AB.   2

    (b) Obtain the determinants of A and of AB.

         Hence, or otherwise, obtain an expression for $\det B$.   3

6. Find the Maclaurin series for $\cos x$ as far as the term in x^4.

    Deduce the Maclaurin series for $f(x) = \dfrac{1}{2}\cos 2x$ as far as the term in x^4.

    Hence write down the first three non-zero terms of the series for $f(3x)$.   5

7. For all natural numbers n, prove whether the following results are true or false.

    (a) $n^3 - n$ is always divisible by 6.   3

    (b) $n^3 + n + 5$ is always prime.   2

MARKS

8. Use integration by parts to obtain $\int 8x^2 \sin 4x\, dx$.   5

9. Obtain the general solution of the equation $\dfrac{d^2y}{dx^2} + 6\dfrac{dy}{dx} + 9y = e^{2x}$.   6

10. Use the substitution $u = 1 + x^2$ to obtain $\displaystyle\int_0^1 \dfrac{x^3}{(1+x^2)^4}\, dx$.

A solid is formed by rotating the curve $y = \dfrac{x^{\frac{3}{2}}}{(1+x^2)^2}$ between $x = 0$ and $x = 1$ through

360° about the x-axis. Write down the volume of this solid.   6

11. The curve $y = x^{2x^2+1}$ is defined for $x > 0$. Obtain the values of y and $\dfrac{dy}{dx}$ at the point where $x = 1$.   5

12. Prove by induction that for $a > 0$,

   $(1 + a)^n \geq 1 + na$

   for all positive integers n.   5

13. The function $f(x)$ is defined by

   $$f(x) = \dfrac{x^2 + 2x}{x^2 - a^2} \qquad (x \neq \pm a).$$

   Obtain equations for the asymptotes of the graph of $f(x)$.

   Show that $f(x)$ is a strictly decreasing function.

   Find the coordinates of the points where the graph of $f(x)$ crosses:

   (i)   the x-axis and

   (ii)   the horizontal asymptote.

   Sketch the graph of $f(x)$, showing clearly all relevant features.   10

MARKS

14. Lines L_1 and L_2 are given by the parametric equations

L_1: $x = 2 + s$, $y = -s$, $z = 2 - s$

L_2: $x = -1 - 2t$, $y = t$, $z = 2 + 3t$.

(a) Show that L_1 and L_2 do not intersect.   3

(b) The line L_3 passes through the point P (1, 1, 3) and its direction is perpendicular to the directions of both L_1 and L_2. Obtain parametric equations for L_3.   3

(c) Find the coordinates of the point Q where L_3 and L_2 intersect and verify that P lies on L_1.   3

(d) PQ is the shortest distance between the lines L_1 and L_2. Calculate PQ.   1

15. (a) Solve the differential equation

$$(x + 1)\frac{dy}{dx} - 3y = (x + 1)^4$$

given that $y = 16$ when $x = 1$, expressing the answer in the form $y = f(x)$.   6

(b) Hence find the area enclosed by the graphs of $y = f(x)$, $y = (1 - x)^4$ and the x-axis.   4

16. Given $z = \cos\theta + i\sin\theta$, use de Moivre's theorem to write down an expression for z^k in terms of θ, where k is a positive integer.

Hence show that $\dfrac{1}{z^k} = \cos k\theta - i\sin k\theta$.

Deduce expressions for $\cos k\theta$ and $\sin k\theta$ in terms of z.

Show that $\cos^2\theta\sin^2\theta = -\dfrac{1}{16}\left(z^2 - \dfrac{1}{z^2}\right)^2$.

Hence show that $\cos^2\theta\sin^2\theta = a + b\cos 4\theta$, for suitable constants a and b.   10

[END OF MODEL PAPER]

ADVANCED HIGHER FOR CfE │ ANSWER SECTION

SQA AND HODDER GIBSON ADVANCED HIGHER FOR CfE MATHEMATICS 2015

ADVANCED HIGHER FOR CfE MATHEMATICS
SPECIMEN QUESTION PAPER

Question	Expected response (Give one mark for each •)	Max mark	Additional guidance (Illustration of evidence for awarding a mark at each •)
1.	Ans: demonstrate result •1 know and start to use quotient rule •2 complete differentiation •3 simplify numerator	3	•1 $\dfrac{(1+x^2)\times 1 - \dots}{(1-x^2)^2}$ •2 $\dfrac{(1+x^2)\times 1 - 2x(x-1)}{(1+x^2)^2}$ •3 $\dfrac{1+x^2-2x^2+2x}{(1+x^2)^2} = \dfrac{1+2x-x^2}{(1+x^2)^2}$
2.	Ans: 6000 •1 correct substitution into general term •2 simplify •3 identify r and find coefficient	3	•1 $\dbinom{6}{r}(2x)^{6-r}\left(-\dfrac{5}{x^2}\right)^r$ •2 $\dbinom{6}{2}2^{6-r}(-5)^r x^{6-3r}$ •3 $\dbinom{6}{2}(2)^4 (-5)^2 = 6000$

Notes:
1. Accept $\dbinom{6}{6-r}(2x)^{6-r}\left(-\dfrac{5}{x^2}\right)^r$ or correct equivalent for •1.
2. If coefficient is found by expanding the expression, only •3 is available.

Question	Expected response (Give one mark for each •)	Max mark	Additional guidance (Illustration of evidence for awarding a mark at each •)
3.	Ans: $\dfrac{1}{2}\sin^{-1}\left(\dfrac{4x}{3}\right)+c$ •1 evidence of identifying an appropiate method •2 re-write in standard form •3 final answer with constant of integration	3	•1 e.g. identify standard integral $\displaystyle\int \dfrac{1}{\sqrt{a^2-x^2}}dx$ •2 $2\displaystyle\int \dfrac{1}{4\sqrt{\left(\dfrac{3}{4}\right)^2-x^2}}dx$ or equivalent •3 $2\times\dfrac{1}{4}\sin^{-1}\left(\dfrac{4x}{3}\right)+c = \dfrac{1}{2}\sin^{-1}\left(\dfrac{4x}{3}\right)+c$

Note:
For •1 accept any appropriate evidence e.g. using substitution $u = 4x$.

Question			Expected response (Give one mark for each •)	Max mark	Additional guidance (Illustration of evidence for awarding a mark at each •)
4.			Ans: $x = 244, y = -163$	4	
			•[1] start correctly		•[1] $729 = 487 \times 1 + 242$
			•[2] show last non–zero remainder = 1		•[2] $487 = 242 \times 2 + 3$ $242 = 80 \times 3 + 2$ $3 = 2 \times 1 + 1$ $2 = 2 \times 1 + 0, \qquad GCD = 1$
			•[3] evidence of two correct back substitutions using $2 = 242 - 3 \times 80$ or $3 = 487 - 242 \times 2$ or $242 = 729 - 487 \times 1$		•[3] $1 = 3 - 2 \times 1 = 3 - (242 - 80 \times 3) = 81 \times 3 - 242$ $= 81(487 - 2 \times 242) - 242$ $= 81 \times 487 - 163 \times 242$ $= 81 \times 487 - 163(729 - 487)$ $= 244 \times 487 - 163 \times 729$ carefully check for equivalent alternatives
			•[4] values for x and y		•[4] $1 = 487 \times 244 - 729 \times 163$ So, $x = 244, y = -163$
5.			Ans: $\dfrac{x^2 e^{3x}}{3} - \dfrac{2xe^{3x}}{9} + \dfrac{2e^{3x}}{27} + c$	5	
			•[1] evidence of application of integration by parts		•[1] $\left(x^2 \int e^{3x}\, dx - \int \left(\int e^{3x} \cdot \dfrac{d}{dx} x^2\, dx \right) dx \right)$
			•[2] correct choice of u and v'		•[2] $u = x^2 \ v' = e^{3x}$
			•[3] correct first application		•[3] $\dfrac{1}{3} x^2 e^{3x} - \dfrac{2}{3} \int xe^{3x}\, dx$ or equivalent
			•[4] start second application		•[4] $\int xe^{3x}\, dx = \dfrac{xe^{3x}}{3} - \dfrac{e^{3x}}{9}$ or equivalent
			•[5] final answer with constant of integration		•[5] $\dfrac{x^2 e^{3x}}{3} - \dfrac{2xe^{3x}}{9} + \dfrac{2e^{3x}}{27} + c$ or equivalent
6.			Ans: $k = \dfrac{3}{2}, -4$	4	
			•[1] starts process for working out determinant		•[1] $3 \begin{vmatrix} -4 & 2 \\ 0 & 1 \end{vmatrix} - k \begin{vmatrix} 3 & 2 \\ k & 1 \end{vmatrix} + 2 \begin{vmatrix} 3 & -4 \\ k & 0 \end{vmatrix}$
			•[2] completing process correctly		•[2] $-12 - k(3 - 2k) + 8k$
			•[3] simplify and equate to 0		•[3] $2k^2 + 5k - 12 = 0$
			•[4] find values of k		•[4] $k = \dfrac{3}{2}, \quad k = -4$

Note:
Accept answer arrived at through row and column operations.

Question			Expected response (Give one mark for each •)	Max mark	Additional guidance (Illustration of evidence for awarding a mark at each •)
7.			Ans: $\dfrac{dV}{dt} = 600\,\pi\,\text{cm}^3\text{s}^{-1}$	5	
			•^{1} interprets rate of change		•^{1} $\dfrac{dA}{dt} = \dfrac{dA}{dr} \times \dfrac{dr}{dt} = 120\pi$
			•^{2} correct expression for $\dfrac{dA}{dr}$		•^{2} $A = 4\pi r^2,\ \dfrac{dA}{dr} = 8\pi r$
			•^{3} find $\dfrac{dr}{dt}$		•^{3} $\dfrac{dr}{dt} = \dfrac{120\pi}{80\pi} = \dfrac{3}{2}$
			•^{4} correct expression for $\dfrac{dV}{dt}$		•^{4} $\dfrac{dV}{dt} = \dfrac{dV}{dr} \times \dfrac{dr}{dt} = 4\pi r^2 \times \dfrac{3}{2}$
			•^{5} evaluates $\dfrac{dV}{dt}$		•^{5} $\dfrac{dV}{dt} = 4\pi(10)^2 \times \dfrac{3}{2} = 600\,\pi\,\text{cm}^3\,\text{s}^{-1}$
8.	(a)	(i)	Ans: $f(x) = 1 + 3x + \dfrac{9}{2}x^2 + \dfrac{9}{2}x^3 + \ldots$	2	
			•^{1} state Maclaurin expansion for e^{3x} up to e^3		•^{1} $f(x) = 1 + \dfrac{3x}{1!} + \dfrac{(3x)^2}{2!} + \dfrac{(3x)^3}{3!} + \ldots$
			•^{2} correct expansion		•^{2} $f(x) = 1 + 3x + \dfrac{9}{2}x^2 + \dfrac{9}{2}x^3 + \ldots$
		(ii)	Ans: $g(x) = \dfrac{1}{4} - \dfrac{1}{4}x + \dfrac{3}{16}x^2 - \dfrac{1}{8}x^3 + \ldots$	3	
			•^{3} correct differentiation of $g(x)$		•^{3} $g(x) = (x+2)^{-2},\ g'(x) = -2(x+2)^{-3},$ $g''(x) = 6(x+2)^{-4},\ g'''(x) = -24(x+2)^{-5}$
			•^{4} correct evaluations of g functions		•^{4} $g(0) = \dfrac{1}{4},\ g'(0) = -\dfrac{1}{4},\ g''(0) = \dfrac{3}{8}$ $g'''(0) = -\dfrac{3}{4}$
			•^{5} correct expansion		•^{5} $g(x) = \dfrac{1}{4} - \dfrac{1}{4}x + \dfrac{3}{16}x^2 - \dfrac{1}{8}x^3 + \ldots$
	(b)		Ans: $h\left(\dfrac{1}{2}\right) = 0.327$	3	
			•^{6} connection between $h(x),\ f(x)$ and $g(x)$		•^{6} $h(x) = x\,f(x)g(x)$
			•^{7} approximate $f\left(\dfrac{1}{2}\right)$ and $g\left(\dfrac{1}{2}\right)$		•^{7} $f\left(\dfrac{1}{2}\right) = 4.1875,\ g\left(\dfrac{1}{2}\right) = 0.15625$
			•^{8} evaluate $h\left(\dfrac{1}{2}\right)$		•^{8} $h\left(\dfrac{1}{2}\right) = 0.327$

Note:
Accept answer given as a fraction.
For (a) (ii) award full credit for answers arrived at using a binomial expansion.
For (b) evidence of use of the expansions from (a) must be evident. Candidates who simply calculate the value of $h\left(\tfrac{1}{2}\right)$ directly without using the approximations from (a) receive no marks for (b).

Question			Expected response (Give one mark for each •)	Max mark	Additional guidance (Illustration of evidence for awarding a mark at each •)
9.			Ans: proof, $r = \dfrac{9}{4}$ •^{1} create term formulae •^{2} form ratios for r •^{3} complete proof •^{4} evaluate r	4	•^{1} $\;u_3 = a + 2d$ $\quad u_7 = a + 6d$ $\quad u_{16} = a + 15d$ •^{2} $\;\dfrac{a+15d}{a+6d} = \dfrac{a+6d}{a+2d}$ •^{3} $\;(a+15d)(a+2d) = (a+6d)^2$ $\quad a^2 + 17ad + 30d^2 = a^2 + 12ad + 36d^2$ $\quad 5a = 6d$ $\quad a = \dfrac{6}{5}d$ •^{4} $\;r = \dfrac{\frac{6}{5}d + 6d}{\frac{6}{5}d + 2d} = \dfrac{9}{4}$
10.			Ans: $\dfrac{dy}{dx} = \dfrac{3}{3x+2} + 2 - \dfrac{4}{2x-1}$ •^{1} introduction of $\log_e$ •^{2} express function in differentiable form •^{3} differentiate	3	•^{1} $\;y = \ln\left(\dfrac{(3x+2)\,e^{2x}}{(2x-1)^2}\right)$ •^{2} $\;y = \ln|3x+2| + 2x - 2\ln|2x-1|$ •^{3} $\;\dfrac{dy}{dx} = \dfrac{3}{3x+2} + 2 - \dfrac{4}{2x-1}$

Note:
In this question the use of modulus signs is not required for the award of •^{1} and •^{2}.

Question			Expected response (Give one mark for each •)	Max mark	Additional guidance (Illustration of evidence for awarding a mark at each •)
11.			Ans: $\ln 2 - \dfrac{1}{6}$ •^{1} correct form of partial fractions •^{2} 1st coefficient correct •^{3} 2nd coefficient correct •^{4} 3rd coefficient correct •^{5} •^{6} integrate any two terms $\qquad$ integrate all three terms •^{7} evaluate	7	•^{1} $\;\dfrac{A}{x+1} + \dfrac{B}{(x+1)^2} + \dfrac{C}{2x-1}$ •^{2} $\;A = -1$ •^{3} $\;B = -1$ •^{4} $\;C = 2$ •^{5} •^{6} $\displaystyle\int_1^2 \left(\dfrac{2}{2x-1} - \dfrac{1}{x+1} - \dfrac{1}{(x+1)^2}\right)dx$ $\qquad = \left[\ln|2x-1| - \ln|x+1| + (x+1)^{-1}\right]_1^2$ •^{7} $\;\ln 2 - \dfrac{1}{6}$

Note:
Do not penalise the omission of the modulus signs at •^{5} and •^{6}.

Question			Expected response (Give one mark for each •)	Max mark	Additional guidance (Illustration of evidence for awarding a mark at each •)
12.	(a)		Ans: m is odd and n is odd •^{1} correct statement	1	•^{1} m is odd and n is odd
	(b)		Ans: proof •^{2} contrapositive statement •^{3} begin proof •^{4} complete proof	3	•^{2} If m and n are both odd then mn is odd. •^{3} Let $m = 2p - 1$, $n = 2q - 1$ where p, q are positive integers. Then, $mn = 2(2pq - p - q) + 1$ where $2pq - p - q$ is clearly an integer therefore mn is clearly odd. •^{4} And so the contrapositive statement is true and it follows that the original statement, 'if mn is even then m is even or n is even', that is equivalent to the contrapositive, is true.

Note:
For •^{1} accept an equivalent statement, e.g. '**neither m nor n is even**' but do not accept any other answer, e.g. 'It is not true to say that m is even or n is even'.

Question			Expected response	Max mark	Additional guidance
13.	(a)		Ans: $x = 4$, $x = -2$ with explanation •^{1} correct asymptotes •^{2} suitable explanation	2	•^{1} $x^2 - 2x - 8 = 0 \Leftrightarrow x = 4$ or $x = -2$ •^{2} y tends towards $\pm\infty$ as $x \to 4$ and $x \to -2$
	(b)	(i)	Ans: false with explanation •^{3} suitable explanation	1	•^{3} The statement is false because the graph meets the x–axis when $x = \dfrac{3}{4}$.
		(ii)	Ans: proof •^{4} method •^{5} complete proof	2	•^{4} e.g. $f(x) = \dfrac{\dfrac{4}{x} - \dfrac{3}{x^2}}{1 - \dfrac{2}{x} - \dfrac{8}{x^2}}$ •^{5} As $x \to \pm\infty$, $f(x) \to \dfrac{0}{1} = 0$ i.e. the line $y = 0$ is a horizontal asymptote

Note:
For •^{2} accept $4x - 3 \neq 0$ at $x = 4$ or $x = -2$.

Question			Expected response	Max mark	Additional guidance
14.	(a)		Ans: $(3, -2, 8)$ •^{1} write lines in parametric form •^{2} create equations for intersection •^{3} solve a pair of these equations (e.g. the first two) for p and t •^{4} check that the third equation is satisfied •^{5} state coordinates of point of intersection	5	•^{1} $\begin{array}{ll} x = 3t - 6 & x = 4p - 5 \\ y = -t + 1 \text{ and} & y = p - 4 \\ z = 2t + 2 & z = 4p \end{array}$ •^{2} $\begin{array}{l} 4p - 5 = 3t - 6 \\ p - 4 = -t + 1 \\ 4p = 2t + 2 \end{array}$ •^{3} $t = 3$ and $p = 2$ •^{4} e.g. $4(2) = 2(3) + 2$ •^{5} evidence of substitution into third equation and $(3, -2, 8)$

Question			Expected response (Give one mark for each •)	Max mark	Additional guidance (Illustration of evidence for awarding a mark at each •)
	(b)		Ans: $-6x - 4y + 7z = 46$	3	
			$\bullet^6$ use vector product to find normal to the plane		$\bullet^6$ $\begin{vmatrix} \mathbf{i} & \mathbf{j} & \mathbf{k} \\ 3 & -1 & 2 \\ 4 & 1 & 4 \end{vmatrix}$
			$\bullet^7$ evaluate normal vector		$\bullet^7$ $-6\mathbf{i} - 4\mathbf{j} + 7\mathbf{k}$
			$\bullet^8$ form equation of plane		$\bullet^8$ $-6x - 4y + 7z = 46$
	(c)		Ans: $49°$	4	
			$\bullet^9$ select correct vectors		$\bullet^9$ $\begin{pmatrix} -6 \\ -4 \\ 7 \end{pmatrix}$ and $\begin{pmatrix} 2 \\ 4 \\ -1 \end{pmatrix}$
			$\bullet^{10}$ complete calculations of $|a|$, $|b|$ and $a \cdot b$		$\bullet^{10}$ $\left\|\begin{pmatrix} -6 \\ -4 \\ 7 \end{pmatrix}\right\| = \sqrt{101}$, $\left\|\begin{pmatrix} 2 \\ 4 \\ -1 \end{pmatrix}\right\| = \sqrt{21}$ and $\begin{pmatrix} -6 \\ -4 \\ 7 \end{pmatrix} \cdot \begin{pmatrix} 2 \\ 4 \\ -1 \end{pmatrix} = -35$
			$\bullet^{11}$ evaluate acute angle between normal to plane and line		$\bullet^{11}$ $40 \cdot 54°$
			$\bullet^{12}$ calculate angle between line and plane		$\bullet^{12}$ $90° - 40 \cdot 54° = 49 \cdot 46°$
15.	(a)		Ans: $\dfrac{2}{(1-x^2)}$	2	
			$\bullet^1$ express function in differentiable form		$\bullet^1$ $\ln(1+x) - \ln(1-x)$
			$\bullet^2$ complete process		$\bullet^2$ $\dfrac{1}{1+x} + \dfrac{1}{1-x} = \dfrac{2}{(1-x^2)}$
	(b)		Ans: $y = \dfrac{x + e - 2\pi}{e^{\sec x}}$	7	
			$\bullet^3$ express in standard form		$\bullet^3$ $\dfrac{dy}{dx} + y\dfrac{\tan x}{\cos x} = \dfrac{1}{e^{\sec x}}$
			$\bullet^4$ form of integrating factor		$\bullet^4$ $IF = e^{\int \frac{\tan x}{\cos x} dx}$
			$\bullet^5$ find integrating factor		$\bullet^5$ $IF = e^{\sec x}$
			$\bullet^6$ state modified equation		$\bullet^6$ $\dfrac{d}{dx}(ye^{\sec x}) = 1$
			$\bullet^7$ integrate both sides		$\bullet^7$ $e^{\sec x} y = x + c$
			$\bullet^8$ substitute in for x and y and find c		$\bullet^8$ $e^{\sec 2\pi} \cdot 1 = 2\pi + c$, $c = e - 2\pi$
			$\bullet^9$ state particular solution		$\bullet^9$ $y = \dfrac{x + e - 2\pi}{e^{\sec x}}$

Question			Expected response (Give one mark for each •)	Max mark	Additional guidance (Illustration of evidence for awarding a mark at each •)
16.	(a)		Ans: proof	5	
			•¹ strategy use partial fractions		•¹ $\dfrac{1}{r(r+1)} = \dfrac{A}{r} + \dfrac{B}{r+1}$
			•² find A and B		•² $A = 1$, $B = -1$
			•³ state result and start to write out series		•³ $1 - \dfrac{1}{2} + \dfrac{1}{2} - \dfrac{1}{3} + \dfrac{1}{3} - \dfrac{1}{4} + \dfrac{1}{4} - \dfrac{1}{5} \cdots$ $+ \dfrac{1}{n-1} - \dfrac{1}{n} + \dfrac{1}{n} - \dfrac{1}{n+1}$
			•⁴ strategy		•⁴ Note that successive terms cancel out (telescopic series) $1 + \left(-\dfrac{1}{2} + \dfrac{1}{2}\right) + \left(-\dfrac{1}{3} + \dfrac{1}{3}\right) + \left(-\dfrac{1}{4} + \dfrac{1}{4}\right)$ $+ \left(-\dfrac{1}{5} + \ldots\right) + \left(\ldots + \dfrac{1}{n-1}\right) + \left(-\dfrac{1}{n} + \dfrac{1}{n}\right) - \dfrac{1}{n+1}$
			•⁵ complete proof		•⁵ cancels terms and $1 - \dfrac{1}{n+1} = \dfrac{n}{n+1}$
	(a)		Ans: proof (alternative)		
			•¹ state hypothesis and consider $n = k + 1$		•¹ Assume $\displaystyle\sum_{r=1}^{k} \dfrac{1}{r(r+1)} = \dfrac{k}{k+1}$ true for some $n = k$, and consider $n = k + 1$ i.e. $\displaystyle\sum_{r=1}^{k+1} \dfrac{1}{r(r+1)} = \sum_{r=1}^{k} \dfrac{1}{r(r+1)} + \dfrac{1}{(k+1)(k+2)}$
			•² start process for $k + 1$		•² $\dfrac{k}{k+1} + \dfrac{1}{(k+1)(k+2)}$ $= \dfrac{k(k+2)}{(k+1)(k+2)} + \dfrac{1}{(k+1)(k+2)}$ $= \dfrac{k^2 + 2k + 1}{(k+1)(k+2)}$
			•³ complete process		•³ $= \dfrac{(k+1)^2}{(k+1)(k+2)}$ $\dfrac{(k+1)}{(k+1)+1}$
			•⁴ show true for $n = 1$		•⁴ For $n = 1$ LHS $= \dfrac{1}{1(1+1)} = \dfrac{1}{2}$ RHS $= \dfrac{1}{1+1} = \dfrac{1}{2}$ LHS = RHS so true for $n = 1$
			•⁵ state conclusion		•⁵ Hence, if true for $n = k$, then true for $n = k + 1$, but since true for $n = 1$, then by induction true for all positive integers n.

Question			Expected response (Give one mark for each •)	Max mark	Additional guidance (Illustration of evidence for awarding a mark at each •)
	(b)	(i)	Ans: $n = 31$	3	
			•6 set up equation and start to solve		•6 e.g. $\dfrac{n+1}{n+2} - \dfrac{n}{n+1} < \dfrac{1}{1000}$ and evidence of strategy
			•7 process		•7 $n^2 + 3n - 998 > 0$
			•8 obtain solution		•8 $n = 31$
		(ii)	Ans: $n = 11$	2	
			•9 set up equation		•9 $\left(\dfrac{n}{n+1}\right)\left(\dfrac{n-1}{n}\right)\left(\dfrac{n-2}{n-1}\right) = \dfrac{n-8}{n-7}$
			•10 solve for n		•10 $n = 11$

Notes:
1   •1 is only available for induction hypothesis and stating that $k + 1$ is going to be considered.
2   •3 is only awarded if final line shows results required in terms of $k + 1$ and is arrived at by appropriate working, including target/desired result approach, from the •3 stage.
3   •5 is only awarded if the candidate shows clear understanding of the logic required.

Question			Expected response (Give one mark for each •)	Max mark	Additional guidance (Illustration of evidence for awarding a mark at each •)
17.	(a)		Ans: proof	5	
			•1 use de Moivre's theorem		•1 $z^4 = \cos 4\theta + i\sin 4\theta$
			•2 start process using binomial theorem		•2 $(\cos\theta + i\sin\theta)^4 = \cos^4\theta + 4\cos^3\theta(i\sin\theta) + 6\cos^2\theta(i\sin\theta)^2 + \ldots$
			•3 complete expansion		•3 $\cos^4\theta + 4\cos^3\theta(i\sin\theta) - 6\cos^2\theta\sin^2\theta + 4\cos\theta(i\sin\theta)^3 + \sin^4\theta$
			•4 identify and match real terms		•4 $\cos 4\theta = \cos^4\theta - 6\cos^2\theta\sin^2\theta + \sin^4\theta$
			•5 identify and match imaginary terms		•5 $\sin 4\theta = 4\cos^3\theta\sin\theta - 4\cos\theta\sin^3\theta$
	(b)		Ans: proof	3	
			•6 strategy		•6 $\tan 4\theta = \dfrac{\sin 4\theta}{\cos 4\theta}$ $= \dfrac{4\cos^3\theta\sin\theta - 4\cos\theta\sin^3\theta}{\cos^4\theta - 6\cos^2\theta\sin^2\theta + \sin^4\theta}$
			•7 divide numerator and denominator by $\cos^4 x$		•7 $\dfrac{\dfrac{4\cos^3\theta\sin\theta}{\cos^4\theta} - \dfrac{4\cos\theta\sin^3\theta}{\cos^4\theta}}{\dfrac{\cos^4\theta}{\cos^4\theta} - \dfrac{6\cos^2\theta\sin^2\theta}{\cos^4\theta} + \dfrac{\sin^4\theta}{\cos^4\theta}}$
			•8 complete		•8 $\dfrac{\dfrac{4\sin\theta}{\cos\theta} - \dfrac{4\sin^3\theta}{\cos^3\theta}}{1 - \dfrac{6\sin^2\theta}{\cos^2\theta} + \dfrac{\sin^4\theta}{\cos^4\theta}} = \dfrac{4\tan\theta - 4\tan^3\theta}{1 - 6\tan^2\theta + \tan^4\theta}$

Question		Expected response (Give one mark for each •)	Max mark	Additional guidance (Illustration of evidence for awarding a mark at each •)
(c)		Ans: $\theta = \dfrac{\pi}{16}$ and $\dfrac{5\pi}{16}$ •^{9} strategy •^{10} complete process and find a solution for 4θ •^{11} find both solutions	3	•^{9} $\tan^4\theta + 4\tan^3\theta - 6\tan^2\theta - 4\tan\theta + 1 = 0$ $4\tan\theta - 4\tan^3\theta = 1 - 6\tan^2\theta + \tan^4\theta$ $\dfrac{4\tan\theta - 4\tan^3\theta}{1 - 6\tan^2\theta + \tan^4\theta} = 1$ •^{10} $\tan 4\theta = 1$, $4\theta = \dfrac{\pi}{4}$ •^{11} $\theta = \dfrac{\pi}{16}$ and $\dfrac{5\pi}{16}$

ADVANCED HIGHER FOR CfE MATHEMATICS
MODEL PAPER 1

Question		Expected response (Give one mark for each •)	Max mark	Additional guidance (Illustration of evidence for awarding a mark at each •)
1.	(a)	Ans: $x = 2$ and $x = -\dfrac{1}{4}$ •^{1} correct use of product rule •^{2} factorise $f'(x)$ •^{3} solve $f'(x) = 0$	3	•^{1} $f'(x) = (x-2)^3 + 3(x+1)(x-2)^2$ •^{2} $f'(x) = (x-2)^2(4x+1)$ •^{3} $x = 2$ and $x = -\dfrac{1}{4}$
	(b)	Ans: $-\dfrac{1}{2}$ **Method 1** •^{4} differentiate LHS of equation •^{5} differentiate RHS of equation •^{6} substitute for x and y or find $\dfrac{dy}{dx}$ •^{7} find gradient of curve **Method 2** •^{4} start to differentiate using quotient rule •^{5} complete differentiation •^{6} substitute for x and y **or** find $\dfrac{dy}{dx}$ •^{7} find gradient of curve	4	•^{4} $2x + x\dfrac{dy}{dx} + y = \dots$ •^{5} $2x + x\dfrac{dy}{dx} + y = 2y\dfrac{dy}{dx} - 5\dfrac{dy}{dx}$ •^{6} $6 + 3\dfrac{dy}{dx} - 1 = -2\dfrac{dy}{dx} - 5\dfrac{dy}{dx}$ **or** $\dfrac{dy}{dx} = \dfrac{2x+y}{2y-x-5}$ •^{7} $5 = -10\dfrac{dy}{dx} \Rightarrow \dfrac{dy}{dx} = -\dfrac{1}{2}$ **or** $\dfrac{dy}{dx} = \dfrac{6-1}{-2-3-5} = \dfrac{5}{-10} = -\dfrac{1}{2}$ •^{4} $\dfrac{2xy - \dots}{y^2}$ •^{5} $\dfrac{2xy - x^2\dfrac{dy}{dx}}{y^2} + 1 = \dfrac{dy}{dx}$ •^{6} $\dfrac{-6 - 9\dfrac{dy}{dx}}{1} + 1 = \dfrac{dy}{dx}$ **or** $\dfrac{dy}{dx} = \dfrac{2xy + y^2}{x^2 + y^2}$ •^{7} $-5 = 10\dfrac{dy}{dx} \Rightarrow \dfrac{dy}{dx} = -\dfrac{1}{2}$ **or** $\dfrac{dy}{dx} = \dfrac{-6+1}{9+1} = \dfrac{-5}{10} = -\dfrac{1}{2}$

Question			Expected response (Give one mark for each •)	Max mark	Additional guidance (Illustration of evidence for awarding a mark at each •)
2.			Ans: 6225 $\bullet^1$ use $u_n = a + (n-1)d$ $\bullet^2$ find d $\bullet^3$ know to use formula for S_n $\bullet^4$ find S_{50}	4	$\bullet^1$ $u_{20} = 97 \Rightarrow a + 19d = 97$ $\bullet^2$ $a = 2 \Rightarrow 2 + 19d = 97 \Rightarrow d = 5$ $\bullet^3$ $S_n = \dfrac{n}{2}[2a + (n-1)d]$ $\bullet^4$ $\dfrac{50}{2}[4 + 49 \times 5] = 6225$
3.			Ans: $3 - 3i$ and -6 $\bullet^1$ show that $3 + 3i$ is a root of the equation $\bullet^2$ find conjugate root $\bullet^3$ find quadratic factor $\bullet^4$ find remaining root	4	$\bullet^1$ $(3+3i)^3 = 27 + 81i + 81i^2 + 27i^3 = -54 + 54i$ $\Rightarrow (3+3i)^3 - 18(3+3i) + 108$ $\qquad = -54 + 54i - 54 - 54i + 108$ $\qquad = 0$ $\bullet^2$ $3 - 3i$ $\bullet^3$ $(z - (3+3i))(z - (3-3i)) = z^2 - 6z + 18$ $\bullet^4$ $z^3 - 18z + 108 = (z^2 - 6z + 18)(z + 6)$ so remaining roots are $3 - 3i$ and -6
4.	(a)		Ans: $x = \pm 2$ $\bullet^1$ find determinant of matrix A $\bullet^2$ solve $\det A = 0$	2	$\bullet^1$ $\det A = 4 - x^2$ $\bullet^2$ $4 - x^2 = 0 \Rightarrow x = \pm 2$
	(b)		Ans: $q = 125$ $\bullet^3$ show that $A^2 = 5A$ $\bullet^4$ begin to express A^4 in terms of A $\bullet^5$ show that $A^4 = 125A$	3	$\bullet^3$ $A^2 = \begin{pmatrix} 1 & 2 \\ 2 & 4 \end{pmatrix}\begin{pmatrix} 1 & 2 \\ 2 & 4 \end{pmatrix} = \begin{pmatrix} 5 & 10 \\ 10 & 20 \end{pmatrix} = 5A$ $\bullet^4$ $A^4 = (A^2)^2 = (5A)^2 = \ldots$ $\bullet^5$ $\ldots (5A)^2 = 25A^2 = 125A \Rightarrow q = 125$
5.	(a)		Ans: $1 + 5x + 10x^2 + 10x^3 + 5x^4 + x^5$ $\bullet^1$ expand $(1 + x)^5$	1	$\bullet^1$ $1 + 5x + 10x^2 + 10x^3 + 5x^4 + x^5$
	(b)		Ans: proof $\bullet^2$ substitute $x = -0.1$ into $(1 + x)^5$ $\bullet^3$ expand $(1 + (-0.1))^5$ and show steps leading to 0.59049	2	$\bullet^2$ $0.9^5 = (1 + (-0.1))^5$ $\bullet^3$ $1 - 0.5 + 0.1 - 0.01 + 0.0005 - 0.00001$ $\qquad = 0.59049$

Question	Expected response (Give one mark for each •)	Max mark	Additional guidance (Illustration of evidence for awarding a mark at each •)
6.	Ans: $\dfrac{3}{8}$ •¹ differentiate $x = 1 + \sin\theta$ •² find limits in terms of x •³ state integral in terms of x •⁴ integrate correctly •⁵ evaluate integral	5	•¹ $x = 1+\sin\theta \;\Rightarrow\; dx = \cos\theta\,d\theta$ •² $\theta = 0 \;\Rightarrow\; x = 1$ $\theta = \dfrac{\pi}{2} \;\Rightarrow\; x = 2$ •³ $\displaystyle\int_1^2 \dfrac{1}{x^3}\,dx$ •⁴ $\left[\dfrac{x^{-2}}{-2}\right]_1^2$ •⁵ $\dfrac{3}{8}$
7.	Ans: $f(x) = 1 + x^2 - \dfrac{x^4}{3}$ •¹ evaluate $f(0)$ •² evaluate $f'(0)$ and $f''(0)$ •³ evaluate $f'''(0)$ and $f''''(0)$ •⁴ state first three terms of $f(x)$	4	•¹ $f(0) = 1$ •² $f'(x) = 2\sin x \cos x = \sin 2x \;\Rightarrow\; f'(0) = 0$ $f''(x) = 2\cos 2x \;\Rightarrow\; f''(0) = 2$ •³ $f'''(x) = -4\sin 2x \;\Rightarrow\; f'''(0) = 0$ $f''''(x) = -8\cos 2x \;\Rightarrow\; f''''(0) = -8$ •⁴ $f(x) = 1 + x^2 - \dfrac{x^4}{3}$
8.	Ans: proof •¹ show true when $n = 1$ •² assume true for $n = k$ •³ consider $n = k+1$ •⁴ simplify •⁵ express in required form and state conclusion	5	•¹ When $n = 1$, LHS $= \dfrac{1}{1\times 2} = \dfrac{1}{2}$, RHS $= 1 - \dfrac{1}{2} = \dfrac{1}{2}$. So true when $n = 1$. •² Assume true for $n = k$, $\displaystyle\sum_{r=1}^{k} \dfrac{1}{r(r+1)} = 1 - \dfrac{1}{k+1}$ •³ Consider $n = k+1$, $\displaystyle\sum_{r=1}^{k+1} \dfrac{1}{r(r+1)} = \sum_{r=1}^{k} \dfrac{1}{r(r+1)} + \dfrac{1}{(k+1)(k+2)}$ •⁴ $= 1 - \dfrac{1}{k+1} + \dfrac{1}{(k+1)(k+2)}$ $= 1 - \dfrac{k+2-1}{(k+1)(k+2)}$ $= 1 - \dfrac{k+1}{(k+1)((k+1)+1)}$ •⁵ $= 1 - \dfrac{1}{((k+1)+1)}$ Thus, if true for $n = k$, statement is true for $n = k+1$, and, since true for $n = 1$, true for all $n \geq 1$.

Question	Expected response (Give one mark for each •)	Max mark	Additional guidance (Illustration of evidence for awarding a mark at each •)
9.	Ans: $y = A\exp\left(2(1+x)^{\frac{3}{2}}\right) - 1$	5	

Method 1

• ¹ separate the variables

$$•^1 \quad \int \frac{dy}{1+y} = 3\int (1+x)^{\frac{1}{2}}\,dx$$

• ² integrate term in y

$$•^2 \quad \ln(1+y) = \ldots$$

• ³ integrate term in x

$$•^3 \quad \ln(1+y) = 2(1+x)^{\frac{3}{2}} + \ldots$$

• ⁴ insert constant and eliminate ln

$$•^4 \quad 1+y = \exp\left(2(1+x)^{\frac{3}{2}} + c\right)$$

• ⁵ solve for y

$$•^5 \quad y = \exp\left(2(1+x)^{\frac{3}{2}} + c\right) - 1$$
$$= A\exp\left(2(1+x)^{\frac{3}{2}}\right) - 1$$

Method 2

• ¹ express in form $\dfrac{dy}{dx} + P(x)y = Q(x)$

$$•^1 \quad \frac{dy}{dx} - 3\left(\sqrt{1+x}\right)y = 3\sqrt{1+x}$$

• ² find integrating factor

$$•^2 \quad \exp\left(-3\int \sqrt{1+x}\,dx\right) = \exp\left(-2(1+x)^{\frac{3}{2}}\right)$$

• ³ express as derivative of product

$$•^3 \quad \frac{d}{dx}\,y\exp\left(-2(1+x)^{\frac{3}{2}}\right)$$
$$= 3\sqrt{1+x}\,\exp\left(-2(1+x)^{\frac{3}{2}}\right)$$

• ⁴ integrate

$$•^4 \quad y\exp\left(-2(1+x)^{\frac{3}{2}}\right)$$
$$= -\int \left(-3\sqrt{1+x}\right)\exp\left(-2(1+x)^{\frac{3}{2}}\right)dx$$
$$= -\exp\left(-2(1+x)^{\frac{3}{2}}\right) + c$$

• ⁵ solve for y

$$•^5 \quad y = -1 + c\exp\left(2(1+x)^{\frac{3}{2}}\right)$$

Question			Expected response (Give one mark for each •)	Max mark	Additional guidance (Illustration of evidence for awarding a mark at each •)
10.			Ans: $\dfrac{\pi}{8} - \dfrac{1}{4}\ln 2$	5	
			•^{1} start to integrate by parts		•^{1} $\left[\tan^{-1}x^2 \int x\ dx\right]_0^1 \cdots$
			•^{2} continue to integrate by parts		•^{2} $\cdots - \dfrac{1}{2}\displaystyle\int_0^1 \dfrac{x^3}{1+x^4}\,dx$
			•^{3} integrate correctly		•^{3} $\left[\dfrac{1}{2}x^2\tan^{-1}x^2\right]_0^1 - \left[\dfrac{1}{4}\ln(1+x^4)\right]_0^1$
			•^{4} substitute limits		•^{4} $\dfrac{1}{2}\tan^{-1}1 - 0 - \left(\dfrac{1}{4}\ln 2 - \dfrac{1}{4}\ln 1\right)$
			•^{5} evaluate integral		•^{5} $\dfrac{\pi}{8} - \dfrac{1}{4}\ln 2$
11.	(a)		Ans: 32	1	
			•^{1} find acceleration when $t=0$		•^{1} $a = \dfrac{dv}{dt} = 3t^2 - 24t + 32$ $\Rightarrow$ when $t=0$, $a=32$
	(b)		Ans: $s = \dfrac{1}{4}t^4 - 4t^3 + 16t^2$; $t=8$	4	
			•^{2} integrate to find general formula for displacement		•^{2} $s = \displaystyle\int t^3 - 12t^2 + 32t\ dt = \dfrac{1}{4}t^4 - 4t^3 + 16t^2 + c$
			•^{3} find formula for displacement when $t=0$		•^{3} $s=0$ when $t=0 \Rightarrow c=0$ $\Rightarrow s = \dfrac{1}{4}t^4 - 4t^3 + 16t^2$
			•^{4} set displacement equal to 0 and factorise		•^{4} at O, $s=0 \Rightarrow \dfrac{1}{4}t^4 - 4t^3 + 16t^2 = 0$ $\Leftrightarrow \dfrac{1}{4}t^2(t^2 - 16t + 64) = 0$ $\Leftrightarrow \dfrac{1}{4}t^2(t-8)^2 = 0$
			•^{5} state time that body returns to O		•^{5} the body returns to O when $t=8$

Question	Expected response (Give one mark for each •)	Max mark	Additional guidance (Illustration of evidence for awarding a mark at each •)
12.	Ans: $a = 4$, $b = 2$, $c = -3$; diagram •[1] substitute $z = x + iy$ into given equation •[2] find expressions for the square of each modulus •[3] rearrange into the form $ax + by + c = 0$ •[4] show locus on Argand diagram	4	•[1] $\|z - 2\| = \|z + i\|$ $\Leftrightarrow \|(x - 2) + iy\| = \|x + (y + 1)i\|$ •[2] $(x - 2)^2 + y^2 = x^2 + (y + 1)^2$ •[3] $x^2 - 4x + 4 + y^2 = x^2 + y^2 + 2y + 1$ $\Leftrightarrow 4x + 2y - 3 = 0$ •[4]
13.	Ans: proof •[1] state assumption that $2 + x$ is rational •[2] express $2 + x$ as a rational number •[3] rearrange to express x as a rational number •[4] state conclusion	4	•[1] Assume $2 + x$ is rational •[2] and let $2 + x = \dfrac{p}{q}$ where p, q are integers. •[3] So $x = \dfrac{p}{q} - 2 = \dfrac{p - 2q}{q}$. •[4] Since $p - 2q$ and q are integers, it follows that x is rational. This is a contradiction.

Question	Expected response (Give one mark for each •)	Max mark	Additional guidance (Illustration of evidence for awarding a mark at each •)
14.	Ans: $y = -4e^x + e^{2x} + x^2 + 3x + \dfrac{7}{2}$ •¹ state the auxiliary equation •² solve the auxiliary equation •³ state the complementary function •⁴ use correct form of particular integral •⁵ substitute for $\dfrac{d^2y}{dx^2}$, $\dfrac{dy}{dx}$ and y in the differential equation •⁶ find values of a, b and c •⁷ state general solution •⁸ obtain equation in A and B by substituting $x = 0$, $y = \dfrac{1}{2}$ and $\dfrac{dy}{dx} = 1$ into the general solution •⁹ obtain equation in A and B by substituting $x = 0$, $y = \dfrac{1}{2}$ and $\dfrac{dy}{dx} = 1$ into the derivative of the general solution •¹⁰ state particular solution	**10**	•¹ $m^2 - 3m + 2 = 0$ •² $(m-1)(m-2) = 0 \Rightarrow m = 1$ or $m = 2$ •³ $y = Ae^x + Be^{2x}$ •⁴ $y = ax^2 + bx + c$ •⁵ $\dfrac{dy}{dx} = 2ax + b$, $\dfrac{d^2y}{dx^2} = 2a$ $\Rightarrow 2a - 3(2ax + b) + 2(ax^2 + bx + c) = 2x^2$ •⁶ $\Leftrightarrow 2ax^2 + (-6a + 2b)x + (2a - 3b + 2c) = 2x^2$ $\Rightarrow a = 1, \ b = 3, \ c = \dfrac{7}{2}$ •⁷ $y = Ae^x + Be^{2x} + x^2 + 3x + \dfrac{7}{2}$ •⁸ $\dfrac{1}{2} = A + B + \dfrac{7}{2} \Rightarrow A + B = -3$ •⁹ $\dfrac{dy}{dx} = Ae^x + 2Be^{2x} + 2x + 3$ $\Rightarrow 1 = A + 2B + 3$ $\Rightarrow A + 2B = -2$ •¹⁰ $A = -4, \ B = 1$ $\Rightarrow y = -4e^x + e^{2x} + x^2 + 3x + \dfrac{7}{2}$

Question			Expected response (Give one mark for each •)	Max mark	Additional guidance (Illustration of evidence for awarding a mark at each •)
15.			Ans: $\sqrt{2}$	**10**	
			•¹ correct form of partial fractions		•¹ $\dfrac{A}{x}+\dfrac{Bx+C}{x^2+1}$
			•² find value of A		•² $1=A(x^2+1)+(Bx+C)x$ $x=0 \;\Rightarrow\; A=1$
			•³ set up equations in B and C		•³ $x=1 \;\Rightarrow\; 1=2+B+C$ $x=-1 \;\Rightarrow\; 1=2+B-C$
			•⁴ express integral in partial fractions		•⁴ $B=-1,\; C=0$ $\Rightarrow\; I(k)=\displaystyle\int_1^k\left(\dfrac{1}{x}-\dfrac{x}{x^2+1}\right)dx$
			•⁵ integrate first fraction		•⁵ $\displaystyle\int_1^k\dfrac{1}{x}\,dx=[\ln x]_1^k$
			•⁶ integrate second fraction		•⁶ $\dfrac{1}{2}\displaystyle\int_1^k\dfrac{2x}{x^2+1}=\dfrac{1}{2}[\ln(x^2+1)]_1^k$
			•⁷ substitute limits		•⁷ $\ln k-\ln 1-\dfrac{1}{2}\ln(k^2+1)+\dfrac{1}{2}\ln 2$
			•⁸ express integral in required form		•⁸ $\ln\dfrac{k\sqrt{2}}{\sqrt{k^2+1}}$
			•⁹ state expression for $e^{I(k)}$		•⁹ $\dfrac{k\sqrt{2}}{\sqrt{k^2+1}}$
			•¹⁰ evaluate $\displaystyle\lim_{k\to\infty} e^{I(k)}$		•¹⁰ $\dfrac{\sqrt{2}}{\sqrt{1+k^{-2}}}\to\sqrt{2}$ as $k\to\infty$
16.	(a)		Ans: $f'(x)=\dfrac{\ln x-1}{(\ln x)^2}$ $f''(x)=\dfrac{2-\ln x}{x(\ln x)^3}$	**4**	
			•¹ start to use quotient rule to find $f'(x)$		•¹ $\dfrac{1\times\ln x-\dots}{(\ln x)^2}$
			•² find $f'(x)$ in simplest form		•² $\dfrac{1\times\ln x-x\times\dfrac{1}{x}}{(\ln x)^2}=\dfrac{\ln x-1}{(\ln x)^2}$
			•³ use quotient rule to find $f''(x)$		•³ $\dfrac{\dfrac{1}{x}\times(\ln x)^2-(\ln x-1)\times\dfrac{2\ln x}{x}}{(\ln x)^4}$
			•⁴ find $f''(x)$ in simplest form		•⁴ $\dfrac{\ln x-2\ln x+2}{x(\ln x)^3}=\dfrac{2-\ln x}{x(\ln x)^3}$

Question		Expected response (Give one mark for each •)	Max mark	Additional guidance (Illustration of evidence for awarding a mark at each •)
	(b)	Ans: (e, e); minimum turning point.	3	
		•[5] find coordinates of stationary point		•[5] $f'(x) = 0$ when $\ln x = 1$ $\Rightarrow\ x = e$ and $y = e$
		•[6] find sign of f'' at stationary point		•[6] At (e, e), $f''(e) = \dfrac{2-1}{e \times 1^3} > 0$
		•[7] state nature of stationary point		•[7] Hence (e, e) is a minimum turning point.
	(c)	Ans: $\left(e^2, \dfrac{1}{2}e^2\right)$	2	
		•[8] find x coordinate of the point of inflexion		•[8] $f''(x) = 0$ when $\ln x = 2\ \Rightarrow\ x = e^2$
		•[9] state the coordinates of the point of inflexion		•[9] $x = e^2\ \Rightarrow\ y = \dfrac{1}{2}e^2$, so the point of inflexion is $\left(e^2, \dfrac{1}{2}e^2\right)$
17.	(a)	Ans: $x = 3$, $y = -2$, $z = -5$	5	
		•[1] set up augmented matrix		•[1] $\begin{array}{ccc|c} 1 & 1 & -1 & 6 \\ 2 & -3 & 2 & 2 \\ -5 & 2 & \lambda & 1 \end{array}$
		•[2] eliminate the x terms from rows 2 and 3		•[2] $\Rightarrow \begin{array}{ccc|c} 1 & 1 & -1 & 6 \\ 0 & -5 & 4 & -10 \\ 0 & 7 & \lambda-5 & 31 \end{array}$
		•[3] eliminate the y term from row 3		•[3] $\Rightarrow \begin{array}{ccc|c} 1 & 1 & -1 & 6 \\ 0 & -5 & 4 & -10 \\ 0 & 0 & 5\lambda+3 & 17 \end{array}$
		•[4] solve for z		•[4] $z = \dfrac{85}{5\lambda+3}$
		•[5] solve for x and y		•[5] $z = -5$ $-5y - 20 = -10 \Rightarrow y = -2$ $x - 2 + 5 = 6 \Rightarrow x = 3$

Question		Expected response (Give one mark for each •)	Max mark	Additional guidance (Illustration of evidence for awarding a mark at each •)
(b)		Ans: proof	2	
		•[6] eliminate y and z in system of equations		•[6] $\begin{aligned} x + y - z &= 6 \quad (1) \\ 2x - 3y + 2z &= 2 \quad (2) \\ 5x \qquad - z &= 20 \ (2)+3\,(1) \\ 4x - y \qquad &= 14 \ (2)+2\,(1) \end{aligned}$
		•[7] use equations to show that $y = 4t - 14$ and $z = 5t - 20$ given that $x = t$		•[7] $y = 4x - 14$ $z = 5x - 20$ $x = t,\ y = 4t - 14,\ z = 5t - 20$
(c)		Ans: 23·0°	4	
		•[8] know how to find angle between the line and the plane		•[8] evidence of $\cos\theta = \dfrac{l \cdot n}{|l||n|}$ where $\theta =$ the angle between the line and the plane, $l =$ the direction of the line and $n =$ the direction of the normal to the plane
		•[9] find $l \cdot n$, $|l|$ and $|n|$		•[9] $l = i + 4j + 5k$ and $n = -5i + 2j - 4k$ $\Rightarrow \cos\theta = \dfrac{l \cdot n}{|l||n|} = \dfrac{-17}{\sqrt{42}\sqrt{45}}$
		•[10] find angle between the line and the normal to the plane		•[10] 113·0°
		•[11] find acute angle between the line and the plane		•[11] 113·0° − 90° = 23·0°

ADVANCED HIGHER FOR CfE MATHEMATICS
MODEL PAPER 2

Question			Expected response (Give one mark for each •)	Max mark	Additional guidance (Illustration of evidence for awarding a mark at each •)
1.			Ans: $\dfrac{x^4}{16} - \dfrac{3x^3}{2} + \dfrac{27x^2}{2} - 54x + 81$	3	
			•¹ correct powers		•¹ $^4C_0\left(\dfrac{x}{2}\right)^4 + {}^4C_1\left(\dfrac{x}{2}\right)^3(-3) + {}^4C_2\left(\dfrac{x}{2}\right)^2(-3)^2$ $+ {}^4C_3\left(\dfrac{x}{2}\right)(-3)^3 + {}^4C_4(-3)^4$
			•² correct coefficients		•² $\left(\dfrac{x}{2}\right)^4 + 4\left(\dfrac{x}{2}\right)^3(-3) + 6\left(\dfrac{x}{2}\right)^2(-3)^2$ $+ 4\left(\dfrac{x}{2}\right)(-3)^3 + (-3)^4$
			•³ simplify		•³ $\dfrac{x^4}{16} - \dfrac{3x^3}{2} + \dfrac{27x^2}{2} - 54x + 81$
2.	(a)		Ans: $\dfrac{1}{(2+x)\sqrt{1+x}}$	3	
			•¹ start to differentiate		•¹ $\dfrac{2\dfrac{d}{dx}\left(\sqrt{1+x}\right)}{1+(1+x)}$
			•² complete differentiation		•² $\dfrac{2 \times \dfrac{1}{2}(1+x)^{-\frac{1}{2}}}{1+(1+x)}$
			•³ simplify		•³ $\dfrac{1}{(2+x)\sqrt{1+x}}$
	(b)		Ans: $\dfrac{-\ln x}{3x^2}$	3	
			•⁴ start to differentiate using quotient rule		•⁴ $\dfrac{\dfrac{1}{x}3x\ldots}{(3x)^2}$
			•⁵ complete differentiation		•⁵ $\dfrac{\dfrac{1}{x}3x - (1+\ln x)3}{(3x)^2}$
			•⁶ simplify		•⁶ $\dfrac{-\ln x}{3x^2}$

Question	Expected response (Give one mark for each •)	Max mark	Additional guidance (Illustration of evidence for awarding a mark at each •)
3.	Ans: $r = \dfrac{1}{2}$; $n = 9$	5	
	•1 use $u_n = ar^{n-1}$		•1 $a = 2048$ and $ar^3 = 256$
	•2 find r		•2 $\Rightarrow r^3 = \dfrac{1}{8} \Rightarrow r = \dfrac{1}{2}$
	•3 set up equation using S_n formula		•3 $\dfrac{2048\left[1-\left(\dfrac{1}{2}\right)^n\right]}{1-\left(\dfrac{1}{2}\right)} = 4088$
	•4 solve for $\left(\dfrac{1}{2}\right)^n$		•4 $1-\left(\dfrac{1}{2}\right)^n = \dfrac{1}{2} \times \dfrac{4088}{2048} = \dfrac{511}{512}$ $\Rightarrow \left(\dfrac{1}{2}\right)^n = 1 - \dfrac{511}{512} = \dfrac{1}{512}$
	•5 solve for n		•5 $2^n = 512 \Rightarrow n = 9$
4.	Ans; $x = +2$	4	
	Method 1		
	•1 recognise that integral is of the form $\int \dfrac{f'}{f}\,dx$		•1 $\dfrac{d}{dx}(e^x - e^{-x}) = e^x + e^{-x}$ (may be implied by •2)
	•2 integrate correctly		•2 $\left[\ln(e^x - e^{-x})\right]_{\ln\frac{3}{2}}^{\ln 2}$
	•3 substitute limits		•3 $\left(\ln 2 - \ln\dfrac{1}{2}\right) - \left(\ln\dfrac{3}{2} - \ln\dfrac{2}{3}\right)$
	•4 show steps leading to integral in required form		•4 $\ln\dfrac{3}{2} - \ln\dfrac{5}{6} = \ln\left(\dfrac{3}{2} \times \dfrac{6}{5}\right) = \ln\dfrac{9}{5}$
	Method 2		
	•1 use integration by substitution		•1 Let $u = e^x - e^{-x}$, then $du = (e^x + e^{-x})dx$
	•2 find limits in terms of u		•2 $x = \ln 2 \Rightarrow u = \dfrac{3}{2}$ $x = \ln\dfrac{3}{2} \Rightarrow u = \dfrac{5}{6}$
	•3 integrate correctly		•3 $\int_{\frac{5}{6}}^{\frac{3}{2}} \dfrac{du}{u} = \left[\ln u\right]_{\frac{5}{6}}^{\frac{3}{2}}$
	•4 show steps leading to integral in required form		•4 $\ln\dfrac{3}{2} - \ln\dfrac{5}{6} = \ln\left(\dfrac{3}{2} \times \dfrac{6}{5}\right) = \ln\dfrac{9}{5}$

Question			Expected response (Give one mark for each •)	Max mark	Additional guidance (Illustration of evidence for awarding a mark at each •)
5.			Ans: $-\dfrac{1}{2}+\dfrac{1}{2}i$; $|z|=\dfrac{1}{2}\sqrt{2}$; $\arg(z)=\dfrac{3\pi}{4}$ or $135°$	6	
			•¹ expand brackets on numerator		•¹ $\dfrac{1+4i-4}{7-i}$
			•² multiply by numerator and denominator by complex conjugate of $7-i$		•² $\dfrac{-3+4i}{7-i}\times\dfrac{7+i}{7+i}$
			•³ express z in the form $a+bi$		•³ $-\dfrac{1}{2}+\dfrac{1}{2}i$
			•⁴ show z on Argand diagram		•⁴
			•⁵ evaluate $|z|$		•⁵ $|z|=\sqrt{\dfrac{1}{4}+\dfrac{1}{4}}=\dfrac{1}{2}\sqrt{2}$
			•⁶ evaluate $\arg(z)$		•⁶ $\arg(z)=\tan^{-1}(-1)=\dfrac{3\pi}{4}$ or $135°$
6.			Ans: $\dfrac{dy}{dx}=\dfrac{-y^2-6xy}{2xy+3x^2}$; $7x+5y=12$	6	
			•¹ differentiate first term		•¹ $y^2+2xy\dfrac{dy}{dx}\ldots$
			•² complete differentiation		•² $y^2+2xy\dfrac{dy}{dx}+6xy+3x^2\dfrac{dy}{dx}=0$
			•³ find $\dfrac{dy}{dx}$		•³ $\dfrac{dy}{dx}=\dfrac{-y^2-6xy}{2xy+3x^2}$
			•⁴ find y when $x=1$		•⁴ $\Rightarrow y^2+3y=4$ $\Rightarrow y^2+3y-4=0$ $\Rightarrow (y+4)(y-1)=0$ $\Rightarrow y=1$ since $y>0$
			•⁵ find gradient of tangent		•⁵ at (1, 1), $\dfrac{dy}{dx}=\dfrac{-7}{5}$
			•⁶ find equation of tangent		•⁶ $(y-1)=-\dfrac{7}{5}(x-1)$ $\Leftrightarrow 7x+5y=12$

Question	Expected response (Give one mark for each •)	Max mark	Additional guidance (Illustration of evidence for awarding a mark at each •)
7.	Ans: $\dfrac{2}{3}$	5	
	•¹ differentiate $u = 1 + x$		•¹ $u = 1 + x \Rightarrow du = dx$
	•² find limits in terms of u		•² $x = 0 \Rightarrow u = 1$ $x = 3 \Rightarrow u = 4$
	•³ state integral in terms of u		•³ $\displaystyle\int_1^4 \dfrac{u-1}{u^{\frac{1}{2}}}\,du$
	•⁴ integrate correctly		•⁴ $\displaystyle\int_1^4 u^{\frac{1}{2}} - u^{-\frac{1}{2}}\,du = \left[\dfrac{u^{\frac{3}{2}}}{\frac{3}{2}} - \dfrac{u^{\frac{1}{2}}}{\frac{1}{2}} \right]_1^4$
	•⁵ evaluate integral		•⁵ $\left(\dfrac{2 \times 8}{3} - 2 \times 2 \right) - \left(\dfrac{2}{3} - 2 \right) = 2\dfrac{2}{3}$
8.	Ans: $x = -\dfrac{1}{3}$, $y = 0$, $z = \dfrac{2}{3}$ When $\lambda = 2$, the second and third rows of the second matrix are the same, so there is an infinite number of solutions.	6	
	•¹ set up augmented matrix		•¹ $\begin{array}{ccc|c} 1 & 1 & 2 & 1 \\ 2 & \lambda & 1 & 0 \\ 3 & 3 & 9 & 5 \end{array}$
	•² eliminate the x terms from rows 2 and 3 and the y term from row 3		•² $\begin{array}{ccc|c} 1 & 1 & 2 & 1 \\ 0 & \lambda-2 & -3 & -2 \\ 0 & 0 & 3 & 2 \end{array}$
	•³ solve for z		•³ $z = \dfrac{2}{3}$
	•⁴ solve for x and y		•⁴ $y = 0, x = -\dfrac{1}{3}$
	•⁵ begin to explain what happens when $\lambda = 2$		•⁵ When $\lambda = 2$, the second and third rows of the second matrix are the same,
	•⁶ complete explanation		•⁶ so there is an infinite number of solutions.

Question	Expected response (Give one mark for each •)	Max mark	Additional guidance (Illustration of evidence for awarding a mark at each •)
9.	Ans: proof; $p = \dfrac{1}{2}$, $q = \dfrac{1}{2}$	6	
	$\bullet^1$ know how to find A^2		$\bullet^1$ $\begin{pmatrix} 0 & 4 & 2 \\ 1 & 0 & 1 \\ -1 & -2 & -3 \end{pmatrix}\begin{pmatrix} 0 & 4 & 2 \\ 1 & 0 & 1 \\ -1 & -2 & -3 \end{pmatrix}$
	$\bullet^2$ find A^2		$\bullet^2$ $\begin{pmatrix} 2 & -4 & -2 \\ -1 & 2 & -1 \\ 1 & 2 & 5 \end{pmatrix}$
	$\bullet^3$ know how to find $A^2 + A$		$\bullet^3$ $\begin{pmatrix} 2 & -4 & -2 \\ -1 & 2 & -1 \\ 1 & 2 & 5 \end{pmatrix} + \begin{pmatrix} 0 & 4 & 2 \\ 1 & 0 & 1 \\ -1 & -2 & -3 \end{pmatrix}$
	$\bullet^4$ find $A^2 + A$ and show that $A^2 + A = 2I$		$\bullet^4$ $= \begin{pmatrix} 2 & 0 & 0 \\ 0 & 2 & 0 \\ 0 & 0 & 2 \end{pmatrix} = 2I$
	$\bullet^5$ multiply $A^2 + A = 2I$ by A^{-1}		$\bullet^5$ $A^2 + A = 2I \;\Leftrightarrow\; A^{-1}(A^2 + A) = 2A^{-1}$
	$\bullet^6$ express A^{-1} in the form $pA + qI$		$\bullet^6$ $A + I = 2A^{-1} \;\Rightarrow\; A^{-1} = \dfrac{1}{2}A + \dfrac{1}{2}I$
10.	Ans: proof; $2q - 12q^2$; proof	5	
	$\bullet^1$ correct method		$\bullet^1$ $\displaystyle\sum_{r=1}^{n} 4 - 6\sum_{r=1}^{n} r = 4n - 6 \times \dfrac{1}{2}n(n+1)$
	$\bullet^2$ complete proof		$\bullet^2$ $4n - 3n^2 - 3n = n - 3n^2$
	$\bullet^3$ write down formula for $\displaystyle\sum_{r=1}^{2q}(4 - 6r)$		$\bullet^3$ $2q - 12q^2$
	$\bullet^4$ correct method		$\bullet^4$ $\displaystyle\sum_{r=1}^{2q}(4 - 6r) - \sum_{r=1}^{q}(4 - 6r)$
	$\bullet^5$ complete proof		$\bullet^5$ $(2q - 12q^2) - (q - 3q^2) = q - 9q^2$
11.	Ans: $60°C$	4	
	$\bullet^1$ differentiate correctly		$\bullet^1$ $\dfrac{dx}{dT} = 3T^2 - 180T + 2400$
	$\bullet^2$ find values of T at stationary points		$\bullet^2$ $3(T - 20)(T - 40) = 0$ at stationary points $\Rightarrow T = 20$ or $T = 40$
	$\bullet^3$ find T at which local maximum occurs		$\bullet^3$ $\dfrac{d^2x}{dT^2} = 6T - 180$ When $T = 20$, $\dfrac{d^2x}{dT^2} < 0$ $\Rightarrow$ local maximum when $T = 20$.
	$\bullet^4$ identify optimal solution by checking values of function at local maximum and at ends of given interval		$\bullet^4$ $x(20) = 20\,000$ $x(10) = 16\,000$ $x(60) = 36\,000$ So the best result is when $T = 60$.

Question			Expected response (Give one mark for each •)	Max mark	Additional guidance (Illustration of evidence for awarding a mark at each •)
12.	(a)		Ans: counter example; proof •¹ state values of m and m^2 for a counter example •² communicate conclusion	2	•¹ e.g. $m = 2$, $m^2 = 4$ •² m^2 is divisible by 4, but m is not divisible by 4, so the statement is false.
	(b)		•³ start proof •⁴ expand brackets •⁵ show that expression represents an odd number and state conclusion	3	•³ Let the numbers be $2n+1$ and $2m$ •⁴ $(2n+1)^3 + (2m)^2$ $= 8n^3 + 12n^2 + 6n + 1 + 4m^2$ •⁵ $= 2(4n^3 + 6n^2 + 3n + 2m^2) + 1$ which is odd.
13.	(a)		Ans: $2x + y = 3$ •¹ find direction vectors of two lines in the plane •² find direction vector of normal to the plane •³ find equation of the plane	3	•¹ $\overrightarrow{AB} = i - 2j$ $\overrightarrow{AC} = -i + 2j + 2k$ •² $\overrightarrow{AB} \times \overrightarrow{AC} = \begin{vmatrix} i & j & k \\ 1 & -2 & 0 \\ -1 & 2 & 2 \end{vmatrix} = -4i - 2j$ •³ $-4x - 2y = -4 \times 1 - 2 \times 1$ $\Leftrightarrow -4x - 2y = -6$ $\Leftrightarrow 2x + y = 3$
	(b)		Ans: $x = 0 + 2t$, $y = 3 - 4t$, $z = 7 - 10t$ •⁴ find a by substituting $(0, a, b)$ into π_1 •⁵ find b and coordinates of point of intersection by substituting $(0, a, b)$ into π_2 •⁶ find direction vector of line of intersection •⁷ find equation of line of intersection	4	•⁴ $2 \times 0 + a = 3 \Rightarrow a = 3$ •⁵ $0 + 3a - b = 2 \Rightarrow b = 3a - 2 = 7$ hence point of intersection is $(0, 3, 7)$. •⁶ $\begin{vmatrix} i & j & k \\ -4 & -2 & 0 \\ 1 & 3 & -1 \end{vmatrix} = 2i - 4j - 10k$ •⁷ $x = 0 + 2t$, $y = 3 - 4t$, $z = 7 - 10t$ or equivalent
	(c)		Ans: $47 \cdot 6°$ •⁸ know how to find angle between the planes •⁹ find $\cos \theta$ •¹⁰ find acute angle between the planes	3	•⁸ evidence of $\cos \theta = \dfrac{n_1 \cdot n_2}{|n_1||n_2|}$ where θ is the angle between the planes; n_1 and n_2 are the direction vectors of the normals to the planes •⁹ $n_1 = -4i - 2j$ and $n_2 = i + 3j - k$ $\Rightarrow \cos \theta = \dfrac{n_1 \cdot n_2}{|n_1||n_2|} = \dfrac{-10}{\sqrt{20}\sqrt{11}}$ •¹⁰ $47 \cdot 6°$

Question	Expected response (Give one mark for each •)	Max mark	Additional guidance (Illustration of evidence for awarding a mark at each •)
14.	Ans: $\dfrac{2}{(x+2)^2} + \dfrac{1}{x-4}$ •^{1} correct form of partial fractions	9	•^{1} $\dfrac{A}{(x+2)^2} + \dfrac{B}{x+2} + \dfrac{C}{x-4}$
	•^{2} find value of A		•^{2} $x^2 + 6x - 4 =$ $A(x-4) + B(x+2)(x-4) + C(x+2)^2$ $x = -2 \Rightarrow A = 2$
	•^{3} find value of C		•^{3} $x = 4 \Rightarrow C = 1$
	•^{4} find value of B and write expression in partial fractions		•^{4} $x = 0 \Rightarrow B = 0$ $\dfrac{2}{(x+2)^2} + \dfrac{1}{x-4}$
	•^{5} evaluate $f(0)$		•^{5} $f(x) = \dfrac{2}{(x+2)^2} + \dfrac{1}{x-4}$ $\Rightarrow f(0) = \dfrac{1}{4}$
	•^{6} differentiate first term		•^{6} $f'(x) = -4(x+2)^{-3} \ldots$
	•^{7} differentiate both terms and evaluate $f'(0)$		•^{7} $f'(x) = -4(x+2)^{-3} - (x-4)^{-2}$ $\Rightarrow f'(0) = -\dfrac{9}{16}$
	•^{8} evaluate $f''(0)$		•^{8} $f''(x) = 12(x+2)^{-4} + 2(x-4)^{-3}$ $\Rightarrow f''(0) = \dfrac{23}{32}$
	•^{9} find first three terms of expansion		•^{9} $\dfrac{1}{4} - \dfrac{9x}{16} + \dfrac{23x^2}{64}$
15. (a)	Ans: $x + y = \sqrt{2}$ •^{1} find $\dfrac{dx}{dt}$	5	•^{1} $\dfrac{dx}{dt} = -2\sin 2t$
	•^{2} find $\dfrac{dy}{dt}$		•^{2} $\dfrac{dy}{dt} = 2\cos 2t$
	•^{3} find $\dfrac{dy}{dx}$		•^{3} $\dfrac{dy}{dx} = \dfrac{2\cos 2t}{-2\sin 2t} = -\cot 2t$
	•^{4} find values of x, y and $\dfrac{dy}{dx}$ when $t = \dfrac{\pi}{8}$		•^{4} if $t = \dfrac{\pi}{8}$, then $x = \dfrac{1}{\sqrt{2}}$, $y = \dfrac{1}{\sqrt{2}}$ and $\dfrac{dy}{dx} = -1$
	•^{5} state equation of tangent		•^{5} $y - \dfrac{1}{\sqrt{2}} = -\left(x - \dfrac{1}{\sqrt{2}}\right) \Leftrightarrow x + y = \sqrt{2}$

Question		Expected response (Give one mark for each •)	Max mark	Additional guidance (Illustration of evidence for awarding a mark at each •)
	(b)	Ans: $\dfrac{d^2y}{dx^2} = \dfrac{-1}{\sin^3 2t}$; $k = -1$	5	
		•6 know how to find $\dfrac{d^2y}{dx^2}$		•6 $\dfrac{d^2y}{dx^2} = \dfrac{\dfrac{d}{dt}\left(\dfrac{dy}{dx}\right)}{\dfrac{dx}{dt}}$
		•7 differentiate correctly		•7 $\dfrac{2\,\text{cosec}^2 2t}{-2\sin 2t}$
		•8 simplify $\dfrac{d^2y}{dx^2}$		•8 $\dfrac{-1}{\sin^3 2t}$
		•9 substitute into LHS of equation		•9 $\sin 2t \dfrac{d^2y}{dx^2} + \left(\dfrac{dy}{dx}\right)^2 = \dfrac{-\sin 2t}{\sin^3 2t} + \left(\dfrac{-\cos 2t}{\sin 2t}\right)^2$
		•10 find value of k		•10 $k = \dfrac{-1 + \cos^2 2t}{\sin^2 2t} = -1$
16.	(a)	Ans: $G = 25k\left(1 - \exp\left(-\dfrac{t}{25}\right)\right)$	4	
		Method 1		
		•1 separate the variables		•1 $\displaystyle\int \dfrac{dG}{25k - G} = \int \dfrac{1}{25}\,dt$
		•2 integrate correctly		•2 $-\ln(25k - G) = \dfrac{t}{25} + C$
		•3 find C		•3 $t = 0$ and $G = 0 \;\rightarrow\; C = \ln 25k$
		•4 express G in terms of t and k		•4 $25k - G = 25k \exp\left(-\dfrac{t}{25}\right)$ $\Rightarrow\; G = 25k\left(1 - \exp\left(-\dfrac{t}{25}\right)\right)$

Question			Expected response (Give one mark for each •)	Max mark	Additional guidance (Illustration of evidence for awarding a mark at each •)
			Method 2		
			$\bullet^1$ express in form $\dfrac{dG}{dt}+P(k)G=Q(k)$		$\bullet^1 \quad \dfrac{dG}{dt}+\dfrac{G}{25}=k$
			$\bullet^2$ integrate correctly		$\bullet^2$ Integrating factor $=$ $\exp\left(\int \dfrac{t}{25}dt\right)=\exp\left(\dfrac{t}{25}\right)$ $\Rightarrow \dfrac{d}{dt}\left(\exp\left(\dfrac{t}{25}\right)G\right)=k\exp\left(\dfrac{t}{25}\right)$ $\Rightarrow \exp\left(\dfrac{t}{25}\right)G=k\int \exp\left(\dfrac{t}{25}\right)dt$ $=k\left(25\exp\left(\dfrac{t}{25}\right)\right)+C$ $\Rightarrow G=25k+C\exp\left(-\dfrac{t}{25}\right)$
			$\bullet^3$ find C		$\bullet^3 \quad t=0$ and $G=0 \Rightarrow C=-25k$
			$\bullet^4$ express G in terms of t and k		$\bullet^4 \quad G=25k\left(1-\exp\left(-\dfrac{t}{25}\right)\right)$
	(b)		Ans: $k\approx 0\cdot 132$ $\bullet^5$ substitute $t=5$ and $G=0.6$ into the equation $\bullet^6$ solve the equation for k correct to three decimal places	2	$\bullet^5 \quad t=5,\ G=0\cdot 6 \Rightarrow 0\cdot 6=25k(1-e^{-0\cdot 2})$ $\bullet^6 \quad k=\dfrac{0\cdot 6}{25(1-e^{-0\cdot 2})}\approx 0\cdot 132$
	(c)		Ans: $G\approx 1\cdot 09$, so the claim seems to be justified. $\bullet^7$ substitute $t=10$ into equation $\bullet^8$ find value of G and state conclusion	2	$\bullet^7 \quad t=10 \Rightarrow G\approx 25\times 0\cdot 132(1-e^{-0\cdot 4})$ $\bullet^8 \quad G\approx 1\cdot 09$, so the claim seems to be justified.
	(d)		Ans: 3·6 metres $\bullet^9$ find the value of G as $t\to\infty$ $\bullet^{10}$ find the long-term height of the plants	2	$\bullet^9 \quad t\to\infty,\ G\to 25k\approx 25\times 0\cdot 132\approx 3\cdot 3$ $\bullet^{10}$ The limit is 3·6 metres.

ADVANCED HIGHER FOR CfE MATHEMATICS
MODEL PAPER 3

Question			Expected response (Give one mark for each •)	Max mark	Additional guidance (Illustration of evidence for awarding a mark at each •)
1.			Ans: $r = -\dfrac{1}{2} \Rightarrow |r| < 1$, so the sum to infinity exists; $S_\infty = 8$	5	
			•^{1} use $u_n = ar^{n-1}$		•^{1} $ar = -6$ and $ar^2 = 3$
			•^{2} find r		•^{2} $r = \dfrac{ar^2}{ar} = \dfrac{3}{-6} = -\dfrac{1}{2}$
			•^{3} state conclusion with justification		•^{3} So, since $|r| < 1$, the sum to infinity exists
			•^{4} correct formula for S_∞		•^{4} $S_\infty = \dfrac{a}{1-r}$
			•^{5} find S_∞		•^{5} $= \dfrac{12}{1-\left(-\dfrac{1}{2}\right)} = 8$
2.	(a)		Ans: $\dfrac{-3}{\sqrt{1-9x^2}}$	2	
			•^{1} start to differentiate		•^{1} $\dfrac{-1}{\sqrt{1-(3x)^2}}$
			•^{2} complete differentiation		•^{2} $\dfrac{-1}{\sqrt{1-(3x)^2}} \times 3 = \dfrac{-3}{\sqrt{1-9x^2}}$
	(b)		Ans: $\dfrac{dy}{dx} = \dfrac{3\cos\theta}{2\sec\theta\tan\theta}$	3	
			•^{3} find $\dfrac{dx}{d\theta}$		•^{3} $\dfrac{dx}{d\theta} = 2\sec\theta\tan\theta$
			•^{4} find $\dfrac{dy}{d\theta}$		•^{4} $\dfrac{dy}{d\theta} = 3\cos\theta$
			•^{5} find $\dfrac{dy}{dx}$		•^{5} $\dfrac{dy}{dx} = \dfrac{3\cos\theta}{2\sec\theta\tan\theta}$
3.			Ans: 7	4	
			•^{1} start to calculate $\mathbf{v} \times \mathbf{w}$		•^{1} $\mathbf{v} \times \mathbf{w} = \begin{vmatrix} \mathbf{i} & \mathbf{j} & \mathbf{k} \\ 3 & 2 & -1 \\ -1 & 1 & 4 \end{vmatrix}$
			•^{2} continue calculating $\mathbf{v} \times \mathbf{w}$		•^{2} $= \mathbf{i}\begin{vmatrix} 2 & -1 \\ 1 & 4 \end{vmatrix} - \mathbf{j}\begin{vmatrix} 3 & -1 \\ -1 & 4 \end{vmatrix} + \mathbf{k}\begin{vmatrix} 3 & 2 \\ -1 & 1 \end{vmatrix}$
			•^{3} calculate $\mathbf{v} \times \mathbf{w}$		•^{3} $= 9\mathbf{i} - 11\mathbf{j} + 5\mathbf{k}$
			•^{4} calculate $\mathbf{u} \cdot (\mathbf{v} \times \mathbf{w})$		•^{4} $-2 \times 9 + 0 \times (-11) + 5 \times 5 = 7$

Question			Expected response (Give one mark for each •)	Max mark	Additional guidance (Illustration of evidence for awarding a mark at each •)
4.			Ans: $34 = 210 \times 1326 - 19 \times 14654$	4	
			•[1] start correctly		•[1] $14654 = 11 \times 1326 + 68$
			•[2] show that gcd = 34		•[2] $1326 = 19 \times 68 + 34$ $68 = 2 \times 34$ So gcd = 34
			•[3] evidence of one correct back substitution		•[3] $34 = 1326 - 19 \times 68$
			•[4] express 34 in the form $1326a + 14654b$		•[4] $34 = 1326 - 19(14654 - 11 \times 1326)$ $= 210 \times 1326 - 19 \times 14654$
5.	(a)		Ans: $\begin{pmatrix} x & x & x \\ -6 & 6 & -1 \\ 0 & 0 & 8 \end{pmatrix}$	2	
			•[1] •[2] eight elements correct nine elements correct		•[1] •[2] $\begin{pmatrix} x & x & x \\ -6 & 6 & -1 \\ 0 & 0 & 8 \end{pmatrix}$
	(b)		Ans: $\det A = 3$, $\det AB = 96x$; $\det B = 32x$	3	
			•[3] find $\det A$		•[3] $\det A = 1 \times (2 + 1) - 0 - 1 \times 0 = 3$
			•[4] find $\det AB$		•[4] $\det AB = x(48 - 0) - x(-48 - 0) + x(0 - 0)$ $= 96x$
			•[5] find $\det B$		•[5] Since $\det AB = \det A \det B$, $\det B = \dfrac{\det AB}{\det A} = \dfrac{96x}{3} = 32x$

Question		Expected response (Give one mark for each •)	Max mark	Additional guidance (Illustration of evidence for awarding a mark at each •)
6.		Ans: $\cos x = 1 - \dfrac{x^2}{2} + \dfrac{x^4}{24}$; $\dfrac{1}{2}\cos 2x = \dfrac{1}{2} - x^2 + \dfrac{x^4}{3}$; $f(3x) = \dfrac{1}{2} - 9x^2 + 27x^4$	5	
		•^{1} find $f(0), f'(0), f''(0),$ $f'''(0)$ and $f''''(0)$		•^{1} $f(0) = \cos 0 = 1$ $f'(x) = -\sin x \implies f'(0) = 0$ $f''(x) = -\cos x \implies f''(0) = -1$ $f'''(x) = \sin x \implies f'(0) = 0$ $f''''(x) = \cos x \implies f'''(0) = 1$
		•^{2} find Maclaurin series for $\cos x$ as far as x^4		•^{2} $1 - \dfrac{x^2}{2} + \dfrac{x^4}{24}$
		•^{3} start to find Maclaurin series for $\dfrac{1}{2}\cos 2x$ as far as x^4		•^{3} $\dfrac{1}{2}\left(1 - \dfrac{(2x)^2}{2} + \dfrac{(2x)^4}{24}\right)$
		•^{4} find Maclaurin series for $\dfrac{1}{2}\cos 2x$ as far as x^4		•^{4} $\dfrac{1}{2} - x^2 + \dfrac{x^4}{3}$
		•^{5} find the first three non-zero terms of the series for $f(3x)$		•^{5} $\dfrac{1}{2} - (3x)^2 + \dfrac{(3x)^4}{3} = \dfrac{1}{2} - 9x^2 + 27x^4$
7.	(a)	Ans: proof	3	
		•^{1} express $n^3 - n$ as a product of three consecutive integers		•^{1} $n^3 - n = n(n^2 - 1) = (n-1)n(n+1)$
		•^{2} start conclusion		•^{2} since $n^3 - n$ is the product of three consecutive integers it is divisible by 3
		•^{3} complete conclusion		•^{3} it is also divisible by 2, so it is divisible by 6
	(b)	Ans: counter example	2	
		•^{4} state values of n and $n^3 + n + 5$ for a counter example		•^{4} e.g. when $n = 2$, $n^3 + n + 5 = 15$
		•^{5} communicate conclusion		•^{5} 15 is not prime, so the statement is false

Question	Expected response (Give one mark for each •)	Max mark	Additional guidance (Illustration of evidence for awarding a mark at each •)
8.	Ans: $-2x^2\cos 4x + x\sin 4x$ $+\dfrac{1}{4}\cos 4x + C$	5	
	•¹ start correct method		•¹ $8x^2\int \sin 4x\,dx \ldots$
	•² continue correct method		•² $\ldots -\int 16x\int(\sin 4x)\,dx$
	•³ integrate correctly		•³ $-2x^2\cos 4x - \int -4x\cos 4x\,dx$
	•⁴ start to integrate $4x\cos 4x$ by parts		•⁴ $-2x^2\cos 4x + 4\left[x\int\cos 4x\,dx - \int\dfrac{1}{4}\sin 4x\,dx\right]$
	•⁵ complete integration		•⁵ $-2x^2\cos 4x + x\sin 4x + \dfrac{1}{4}\cos 4x + C$
9.	Ans: $y=(A+Bx)e^{-3x}+\dfrac{1}{25}e^{-2x}$	6	
	•¹ state the auxiliary equation		•¹ $m^2+6m+9=0$
	•² state the complementary function		•² $(m+3)^2=0 \implies m=-3$ $\implies y=(A+Bx)e^{-3x}$
	•³ use correct form of particular integral		•³ $y=ke^{2x}$
	•⁴ find $\dfrac{dy}{dx}$ and $\dfrac{d^2y}{dx^2}$		•⁴ $\dfrac{dy}{dx}=2ke^{2x}, \quad \dfrac{d^2y}{dx^2}=4ke^{2x}$
	•⁵ find k		•⁵ $4ke^{2x}+12ke^{2x}+9ke^{2x}=e^{2x}$ $\implies 25k=1 \implies k=\dfrac{1}{25}$
	•⁶ state general solution		•⁶ $y=(A+Bx)e^{-3x}+\dfrac{1}{25}e^{-2x}$
10.	Ans: $\dfrac{1}{24}; \dfrac{\pi}{24}$	6	
	•¹ differentiate $u=1+x^2$		•¹ $u=1+x^2 \implies du=2x\,dx$
	•² find limits in terms of u		•² $x=0 \implies u=1, \quad x=1 \implies u=2$
	•³ state integral in terms of u		•³ $\int_1^2 \dfrac{u-1}{2u^4}\,du$
	•⁴ integrate correctly		•⁴ $\dfrac{1}{2}\int_1^2(u^{-3}-u^{-4})\,du = \dfrac{1}{2}\left[-\dfrac{1}{2}u^{-2}+\dfrac{1}{3}u^{-3}\right]_1^2$
	•⁵ evaluate integral		•⁵ $\dfrac{1}{2}\left(-\dfrac{1}{8}+\dfrac{1}{24}\right)-\dfrac{1}{2}\left(-\dfrac{1}{2}+\dfrac{1}{3}\right)=\dfrac{1}{24}$
	•⁶ find volume of solid		•⁶ $\pi\int_0^1 \dfrac{x^3}{(1+x^2)^4}\,dx=\dfrac{\pi}{24}$

Question			Expected response (Give one mark for each •)	Max mark	Additional guidance (Illustration of evidence for awarding a mark at each •)
11.			Ans: $y = 1$, $\dfrac{dy}{dx} = 3$	5	
			•^{1} find y when $x = 1$		•^{1} $x = 1 \Rightarrow y = 1^3 = 1$
			•^{2} take ln of both sides of equation		•^{2} $\ln y = \ln x^{2x^2+1}$ $= (2x^2 + 1)\ln x$
			•^{3} differentiate LHS of equation		•^{3} $\dfrac{1}{y}\dfrac{dy}{dx} = \ldots$
			•^{4} differentiate RHS of equation		•^{4} $\dfrac{1}{y}\dfrac{dy}{dx} = \dfrac{2x^2+1}{x} + 4x\ln x$
			•^{5} find $\dfrac{dy}{dx}$ when $x = 1$ and $y = 1$		•^{5} $x = 1,\ y = 1 \Rightarrow \dfrac{dy}{dx} = 3 + 0 = 3$
12.			Ans: proof	5	
			•^{1} prove true when $n = 1$		•^{1} When $n = 1$, LHS $= 1 + a$ RHS $= 1 + a$ i.e. LHS = RHS So it is true when $n = 1$
			•^{2} assume true for $n = k$		•^{2} Assume true for $n = k$, $(1 + a)^k \geq 1 + ka$
			•^{3} consider $n = k + 1$		•^{3} Consider $n = k + 1$, $(1 + a)^{k+1} = (1 + a)(1 + a)^k$
			•^{4} use assumption for $n = k$		•^{4} $\geq (1 + a)(1 + ka)$ for $a > 0$
			•^{5} express in required form and state conclusion		•^{5} $= 1 + a + ka + ka^2$ $= 1 + (k + 1)a + ka^2$ $> 1 + (k + 1)a$ since $ka^2 > 0$ Thus, if true for $n = k$, statement is true for $n = k + 1$, and, since true for $n = 1$, true for all $n \geq 1$.

Question			Expected response (Give one mark for each •)	Max mark	Additional guidance (Illustration of evidence for awarding a mark at each •)
13.			Ans: asymptotes $x = -a,\ x = a,\ y = 1$; intercepts with (i) x-axis (0, 0) and (−2, 0) (ii) horizontal asymptote $\left(-\dfrac{1}{2}a,\ 1\right)$	10	
			•[1] find equations of vertical asymptotes		•[1] $f(x) = \dfrac{x^2 + 2x}{(x+a)(x-a)}$, so there are vertical asymptotes at $x = -a$ and $x = a$
			•[2] express $f(x)$ in suitable form for finding non-vertical asymptote		•[2] $f(x) = \dfrac{x^2 + 2x}{x^2 - a^2} = 1 + \dfrac{2x + a^2}{x^2 - a^2}$
			•[3] find equation of horizontal asymptote		•[3] $1 + \dfrac{2x + 1}{x^2 - a^2} \to 1$ as $x \to \infty$ so $y = 1$ is a horizontal asymptote
			•[4] find $f'(x)$		•[4] $f'(x) = \dfrac{(2x + 2)(x^2 - a^2) - (x^2 + 2x)2x}{(x^2 - a^2)^2}$
			•[5] express $f'(x)$ in simplest form		•[5] $f'(x) = \dfrac{-2(x^2 + a^2 x + a^2)}{(x^2 - a^2)^2}$
			•[6] show that $f'(x) < 0$ and state conclusion		•[6] $f'(x) = \dfrac{-2\left(\left(x + \dfrac{1}{2}a\right)^2 + \dfrac{3}{4}a^2\right)}{(x^2 - a^2)^2} < 0$ so $f(x)$ is a strictly decreasing function
			•[7] find intercepts with x-axis		•[7] $\dfrac{x(x + 2)}{(x + a)(x - a)} = 0 \ \Rightarrow\ x = 0,\ x = -2$ so graph crosses x-axis at (0, 0) and (−2, 0)
			•[8] find intercept with horizontal asymptote		•[8] $\dfrac{x^2 + 2x}{x^2 - a^2} = 1 \ \Rightarrow\ x^2 + 2x = x^2 - a^2 \ \Rightarrow\ x = -\dfrac{1}{2}a^2$ so graph crosses horizontal asymptote at $\left(-\dfrac{1}{2}a^2, 1\right)$.
			•[9] •[10] sketch curve with correct shape sketch curve in correct position		•[9] •[10]

Question			Expected response (Give one mark for each •)	Max mark	Additional guidance (Illustration of evidence for awarding a mark at each •)
14.	(a)		Ans: proof •^{1} equate the x-coordinates and the y-coordinates •^{2} solve pair of equations to find values of t and s •^{3} complete proof	3	•^{1} $2+s=-1-2t \Rightarrow s+2t=-3$ $-s=t \Rightarrow s=-t$ •^{2} $-t+2t=3 \Rightarrow t=-3;\ s=3$ •^{3} putting $s=3$ in L_1 gives $(5,-3,-1)$ putting $t=-3$ in L_2 gives $(5,-3,-7)$ The z-coordinates differ, so L_1 and L_2 do not intersect
	(b)		Ans: $x=1-2u,\ \ y=1-u,\ \ z=3-u$ •^{4} know how to find direction of L_3 •^{5} find direction of L_3 •^{6} find parametric equations of L_3	3	•^{4} direction of L_3 given by vector product of directions of L_1 and L_2 •^{5} $(i-j-k)\times(-2i+j+3k)$ $=\begin{vmatrix} i & j & k \\ 1 & -1 & -1 \\ -2 & 1 & 3 \end{vmatrix} = -2i-k-j$ •^{6} $r=i+j+3k+(-2i-j-k)u$ $\Rightarrow x=1-2u,\ \ y=1-u,\ \ z=3-u$
	(c)		Ans: $Q(-1,0,2)$ •^{7} equate the x-coordinates and the y-coordinates and solve the pair of equations to find u and t •^{8} find coordinates of Q •^{9} verify that P lies on L_1	3	•^{7} $-1-2t=1-2u \Rightarrow t=1-u$ $-1=3-4u \Rightarrow u=1$ and $t=0$ •^{8} $\Rightarrow x=1-2u=-1,\ y=1-u=0,\ z=3-u=2$ so point of intersection is $Q\ (-1,0,2)$ •^{9} L_1 is $x=2+s,\ y=-s,\ z=2-s$. When $x=1,\ s=-1$ and hence $y=1$ and $z=3$, i.e. P lies on L_1.
	(d)		Ans: $\sqrt{6}$ •^{10} calculate PQ	1	•^{10} $PQ=\sqrt{2^2+1^2+1^2}=\sqrt{6}$

Question			Expected response (Give one mark for each •)	Max mark	Additional guidance (Illustration of evidence for awarding a mark at each •)
15.	(a)		Ans: $y = (x+1)^4$	6	
			•^{1} express in form $\dfrac{dy}{dx} + P(x)y = Q(x)$		•^{1} $\dfrac{dy}{dx} - \dfrac{3}{x+1}y = (x+1)^3$
			•^{2} begin to find integrating factor		•^{2} $\displaystyle\int \dfrac{-3}{x+1}\,dx = -3\ln(x+1)$
			•^{3} find integrating factor		•^{3} $e^{-3\ln(x+1)} = (x+1)^{-3}$
			•^{4} multiply equation by integrating factor		•^{4} $\dfrac{1}{(x+1)^3}\dfrac{dy}{dx} - \dfrac{3}{(x+1)^4}y = 1$
			•^{5} integrate		•^{5} $\dfrac{d}{dx}\left(\dfrac{y}{(x+1)^3}\right) = 1$ $\Rightarrow \dfrac{y}{(x+1)^3} = \displaystyle\int 1\,dx = x + C$
			•^{6} find particular solution		•^{6} $y = 16$ and $x = 1 \Rightarrow \dfrac{16}{2^3} = 1 + C \Rightarrow C = 1$ Hence $y = (x+1)^4$.
	(b)		Ans: $\dfrac{2}{5}$	4	
			•^{7} find x-coordinate of point of intersection curves		•^{7} $(x+1)^4 = (1-x)^4$ $\Leftrightarrow x+1 = 1-x \Rightarrow x = 0$
			•^{8} know how to find area		•^{8} $\displaystyle\int_{-1}^{0}(x+1)^4\,dx + \int_{0}^{1}(1-x)^4\,dx$
			•^{9} integrate correctly		•^{9} $2\displaystyle\int_{-1}^{0}(x+1)^4\,dx = 2\left[\dfrac{(x+1)^5}{5}\right]_{-1}^{0}$
			•^{10} find area		•^{10} $2\left(\dfrac{1}{5} - \dfrac{0}{5}\right) = \dfrac{2}{5}$

Question			Expected response (Give one mark for each •)	Max mark	Additional guidance (Illustration of evidence for awarding a mark at each •)
16.			Ans: $z^k = \cos k\theta + i\sin k\theta$; $\dfrac{1}{z^k} = \cos k\theta - i\sin k\theta$; $\cos k\theta = \dfrac{1}{2}\left(z^k + \dfrac{1}{z^k}\right)$; $\sin k\theta = \dfrac{1}{2i}\left(z^k - \dfrac{1}{z^k}\right)$; $\cos^2\theta\sin^2\theta = \dfrac{1}{8} - \dfrac{1}{8}\cos 4\theta$	10	
			•¹ express z^k in terms of θ		•¹ $z^k = \cos k\theta + i\sin k\theta$
			•² express $\dfrac{1}{z^k}$ in terms of θ		•² $\dfrac{1}{z^k} = z^{-k} = \cos(-k\theta) + i\sin(-k\theta)$
			•³ complete proof		•³ $= \cos k\theta - i\sin k\theta$
			•⁴ express $\cos k\theta$ in terms of z		•⁴ $z^k + \dfrac{1}{z^k} = 2\cos k\theta \Rightarrow \cos k\theta = \dfrac{1}{2}\left(z^k + \dfrac{1}{z^k}\right)$
			•⁵ express $\sin k\theta$ in terms of z		•⁵ $z^k - \dfrac{1}{z^k} = 2i\sin k\theta \Rightarrow \sin k\theta = \dfrac{1}{2i}\left(z^k - \dfrac{1}{z^k}\right)$
			•⁶ start to find expression for $\cos^2\theta\sin^2\theta$		•⁶ For $k-1$, $\cos^2\theta\sin^2\theta = (\cos\theta\sin\theta)^2$
			•⁷ express $\cos^2\theta\sin^2\theta$ in terms of z		•⁷ $= \left(\dfrac{\left(z + \dfrac{1}{z}\right)\left(z - \dfrac{1}{z}\right)}{4i}\right)^2$
			•⁸ complete proof		•⁸ $= \dfrac{\left(z^2 - \dfrac{1}{z^2}\right)^2}{16i^2} = -\dfrac{1}{16}\left(z^2 - \dfrac{1}{z^2}\right)^2$
			•⁹ express $\left(z^2 - \dfrac{1}{z^2}\right)^2$ in terms of θ		•⁹ $\left(z^2 - \dfrac{1}{z^2}\right)^2 = z^4 + \dfrac{1}{z^4} - 2 = 2\cos 4\theta - 2$
			•¹⁰ complete proof		•¹⁰ $\Rightarrow \cos^2\theta\sin^2\theta = -\dfrac{1}{16}(2\cos 4\theta - 2)$ $= \dfrac{1}{8} - \dfrac{1}{8}\cos 4\theta$

Acknowledgements

Hodder Gibson would like to thank the SQA for use of any past exam questions that may have been used in model papers, whether amended or in original form.